Nutrition and Health

Nutrition and Health

Gerald Wiseman MD PhD

Department of Biomedical Science
University of Sheffield
UK

London and New York

First published 2002
by Taylor & Francis
11 New Fetter Lane, London EC4P 4EE

Simultaneously published in the USA and Canada
by Taylor & Francis Inc,
29 West 35th Street, New York, NY 10001

Taylor & Francis is an imprint of the Taylor & Francis Group

Typeset in 10/12pt Baskerville by Graphicraft Limited, Hong Kong
Printed and bound in Great Britain by MPG Books Ltd, Bodmin

British Library Cataloguing in Publication Data
A catalogue record for this book is available from the British Library

Library of Congress Cataloging in Publication Data
A catalog record for this book has been requested

ISBN 0-415-27874-0 (pbk)
ISBN 0-415-27875-9 (hbk)

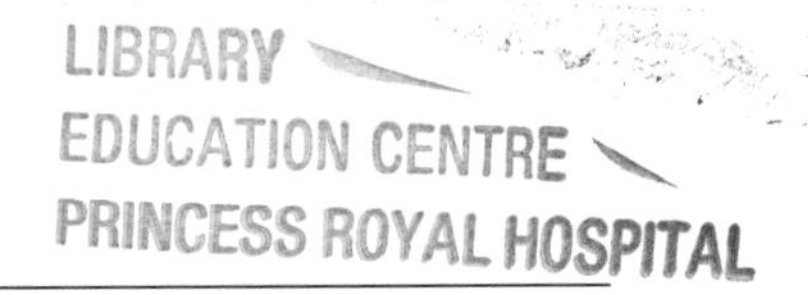

Contents

Tables

Preface

This account of human nutrition describes the basic facts in a clear and simple way without the use of complicated details or much specialist language. In the few places where more than this is necessary, elementary explanations are given. I believe that any averagely intelligent person will readily gain a good knowledge of human nutrition from this book, which will also be of value to students, teachers, nurses, doctors and health professionals.

I would like to thank Professors Anthony Angel and Peter W. Andrews of the Department of Biomedical Science, University of Sheffield, for the very generous facilities they provided during the several years it has taken to write this book.

Gerald Wiseman

Chapter 1

Energy

All the energy needed for growth and repair of the body, for muscular activity of all kinds and for all the work done by cells comes from the metabolism of carbohydrate, fat, protein and alcohol. The numerous other items of the diet, even though essential for other reasons, do not provide energy, although many are directly involved in the chemical reactions which yield energy. If the diet is adequate and properly balanced the energy normally comes chiefly from carbohydrate and fat, while most of the protein is used for cell growth and repair. When there is not enough carbohydrate and fat, the protein is used for energy and is then not available for other purposes. As dietary protein is generally less abundant than carbohydrate and fat, and usually more expensive, using protein for energy is comparatively wasteful. In some communities, however, there may be plentiful protein and it may then be eaten in sufficient quantity to be used for both cell building and for energy.

The intake of food is governed in health by the appetite which under ordinary conditions controls the weight of the body with remarkable precision. Many people taking only moderate care are able to keep their weight more or less unchanged over several decades. If they take food in excess by only a small amount, that excess energy can be disposed of as heat and thereby prevent fat accumulation. This seems to work very efficiently in some people. It is, however, easy to over-ride the natural controlling mechanism and consume substantially more energy than is required. When this happens the excess energy is stored in the body as fat.

During ageing there is a fall in the weight of the bones, due to loss of minerals, plus a fall in the weight of the muscles, hence if the total body weight remains constant there must be compensatory changes, mainly an increase in the body fat.

The ability of the body to override the mechanism which controls energy intake has survival value when the supply of food is unpredictable because it enables fat to be accumulated when there is plenty of food and its energy to be used later when food is scarce. How long a healthy adult can survive without food depends to a large extent on the fat stored: with adequate water, people have lived for many weeks. When people die during starvation they often still have some fat in their body. They die because during starvation body protein is metabolised as well as body fat and it is the loss of the protein that is usually the cause of death. The control of body weight is dealt with in Chapter 2.

Nutritional status

Body mass index (BMI) =
Weight (kg) ÷ Height (m)^2

The nutritional status of most people can be assessed sufficiently well by their appearance, body weight and by simple questions about general health. For a more critical assessment their body mass index can be determined. This gives a weight for height ratio and is a good guide to underweight or overweight in adults except for those who are extremely muscular or have excessive accumulations of water in the body. The use of the body mass index is described in Chapter 2 on obesity.

If weighing is not possible, an assessment can be made by measuring the circumference of the upper arm with a tape-measure. A point midway between the shoulder and the elbow is used with the arm at rest, preferably hanging down. This simple measurement reflects the size of the underlying muscles and the subcutaneous fat, as well as the bone and the skin. In undernourished persons and in those overweight it will be the muscles and the fat which will change in bulk rather than the other tissues. For adult men on a satisfactory diet the circumference ranges from about 250–320 mm and for women from about 220–300 mm.

In children chronic energy lack causes a low height for age ratio, especially if the parents and siblings are of average height or more.

Energy content of food

1 g fat = 9 kcal (38 kjoule)
1 g carbohydrate = 4 kcal (17 kjoule)
1 g protein = 4 kcal (17 kjoule)
1 g alcohol = 7 kcal (29 kjoule)

When carbohydrate, fat and alcohol are metabolised for energy in the body they are normally converted completely to carbon dioxide and water, with energy being released during the process. Protein metabolism during energy release yields various nitrogen-containing substances in addition to carbon dioxide and water. By mimicking these reactions in laboratory experiments the energy value of any food can be measured and expressed as kilocalories (kcal) or kilojoules (kjoule) per gram of the food. One kcal is equal to 4.18 kjoule. The energy values for carbohydrate, fat and protein are approximately 4 kcal (17 kjoule) per gram for carbohydrate and protein and 9 kcal (38 kjoule) per gram for fat. For alcohol, the value is 7 kcal (29 kjoule) per gram. Hence if the amounts of carbohydrate, fat, protein and alcohol in a meal are known, the energy value of the meal can be calculated easily.

Table 1 Energy provided by common foods

Very high (More than 500 kcal/100 g edible portion)	Almonds, Brazil nuts, butter, chocolate, fats, hazelnuts, oils, peanuts, walnuts
High (250–500 kcal/100 g edible portion)	Beef (medium fat), cheese, cornflakes, dates (dried), herring, honey, lamb, mackerel, raisins, rice
Medium (50–250 kcal/100 g edible portion)	Avocado, bananas, beans, beef (lean), bread, chicken (no skin), cod, duck (no skin), egg, figs (dried), haddock, ham (lean), heart, kidney, lentils, liver, maize, milk, peas, pork (lean), potatoes, prawns, prunes, rabbit, salmon, tongue, turkey (no skin), veal
Low (Less than 50 kcal/100 g edible portion)	Apples, apricots (fresh), broccoli, Brussels sprouts, cabbage, carrots, cauliflower, celery, grapefruit, lettuce, mushrooms, oranges, pears, plums, spinach, sweet peppers, tomatoes, watercress

Some foods are energy-rich because they contain little or no water, fibre or other material which does not yield energy; examples are metabolizable sugars, fats and oils. Foods with much water and dietary fibre are usually energy-poor. For example, 100 g table sugar (sucrose) will provide 400 kcal, whereas 100 g of items such as lettuce, tomatoes or cucumber, which contain about 95 per cent water plus fibre, will provide only about 20 kcal. Eating most salad items instead of sugar, fats and oils greatly reduces the energy intake.

The energy values of some everyday foods are shown in Table 1. Natural foods vary in composition from sample to sample and the values given in tables are average ones. This is especially so for animal products in which the fat content may be very variable.

The amount of carbohydrate, protein and fat in the diet varies greatly but an average national picture in 1983 showed that about 12 per cent of the daily calories came from protein, about 46 per cent from carbohydrate and about 42 per cent from fat. From then until 1996 the protein intake was more or less constant, the carbohydrate fell by a few percent, while the fat eaten rose slightly, despite repeated advice that the fat content of the average diet was excessive. A much healthier intake would be about 12 per cent of calories from protein, about 58 per cent from carbohydrate and only 30 per cent from fat. Many children become habituated to eating high-fat foods and as adults they dislike changing their habits. The food industry does not produce on a mass scale a sufficient variety of attractive low-fat foods, particularly snacks.

Energy expenditure

Part of the energy produced by the body may be used for the production of extra tissue during growth or tissue repair and this energy does not appear as heat. It is locked in the new tissue and although it can be estimated it is often ignored. In

contrast, the rest of the metabolic energy, which in adults is virtually all of it, does appear as heat and can be measured accurately in specialized laboratories. The technique is called direct calorimetry. Because this requires special expensive apparatus and is very time consuming, the energy produced by the body can also be calculated from the amount of oxygen taken up and the carbon dioxide given off in the breath. This method, called indirect calorimetry, is easy, cheap and relatively quick. These basic experiments on energy production were first carried out at the end of the nineteenth century and since then many measurements have been made of the energy produced by adults and children while resting or engaged in all sorts of activities. Knowing how much energy is produced each day tells us how much energy needs to be eaten, which enables suitable diets to be designed for all occasions.

The results of these investigations show that almost all normal adults need about 500 kcal for the usual eight hours of sleep. The energy needed for eight hours of work and for eight hours of non-work, however, varies considerably, as would be expected. People who do sedentary work requiring little physical activity need about 2200 kcal per 24 hours, those who do moderately active work need about 2500 kcal per 24 hours, while the few who undertake heavy work require 3000–3500 kcal per 24 hours. Moderately light housework needs only about 2000 kcal per 24 hours but this goes up if there are young children to care for, when the amount of physical activity may be greatly increased.

A person expending about 2200 kcal per 24 hours gives off as much heat as does a lit 100 watt electric light bulb. The skin is not as hot as the bulb because the body has a much larger surface area for heat loss, but in both cases the total heat being lost is about the same.

The values given here and elsewhere for the energy expended during different activities are only guidelines and may vary greatly from subject to subject and often in the same subject doing the same thing at different times.

Effect of body weight

The energy requirement of overweight people is usually less than that of thin people of similar age. This is partly because in the overweight the thicker layer of fat under the skin reduces the body's heat loss, so that less heat production is needed to keep the body temperature normal, requiring less food to be metabolised. In addition, overweight people tend to be less active and therefore need to produce less energy. However, when overweight people are active, the extra weight they carry needs extra energy and their food requirement may go up very markedly.

Effect of age

During the process of ageing the energy requirement gradually decreases. In part this is because in older people some muscle, which is metabolically very active, is often replaced by fat. The ageing process is also accompanied by a fall in the hormones which normally keep the metabolic rate high. Between the ages of about 20 years and 80 years the resting energy requirement falls by an average of about 15 per cent. The total energy requirement of older people may also decrease because many become less mobile, although some do remain remarkably active and may need a higher food intake than some much younger people. The energy intake on ageing needs to be reduced to match any fall in energy expenditure to prevent the familiar gaining of weight with the passage of years, mostly accepted as inevitable, though it does not need to be so.

Effect of exercise

The amount of energy used during exercise is closely related to movement. Frequently moving the body greatly increases the energy used, especially if the body is lifted rather than merely moved horizontally. For using up energy and thereby using up body fat, most forms of exercise are not very good. For example, walking six miles on the level in two hours would be considered an energetic pursuit for many adults yet it uses up only about 500 kcal, which is

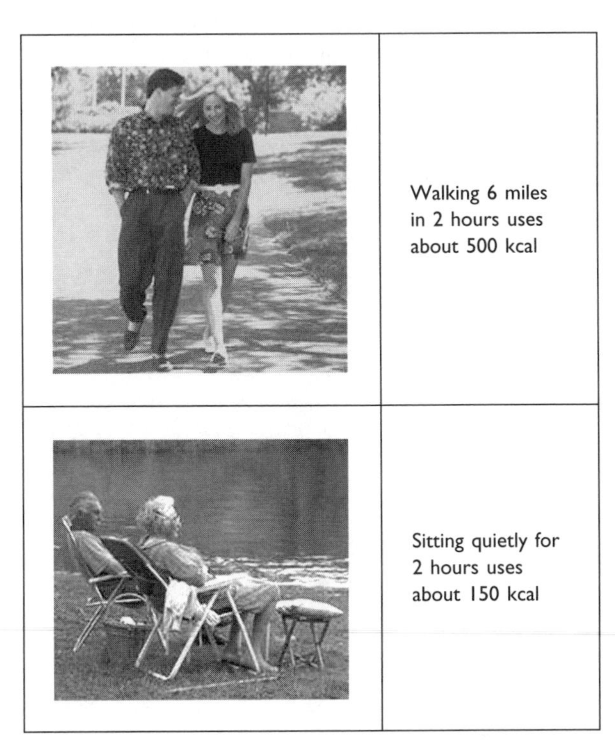

Walking 6 miles in 2 hours uses about 500 kcal

Sitting quietly for 2 hours uses about 150 kcal

about 350 kcal more than simply sitting quietly. Three slices of bread with butter supply about 350 kcal. If all the extra 350 kcal used during this exercise came from the body fat the weight of this lost fat would be about 40 g, representing about 50 g of the adipose tissue which stores the fat. If this six mile walk were undertaken every day for a week, the loss of body weight might be about 350 g (about three-quarters of a pound). Most people will find this a not very impressive result. It may seem much better to eat three slices of bread with butter less each day! Regular exercise is nevertheless very desirable for general fitness.

Effect of undernutrition

When food is plentiful the energy expended is replenished by eating. This is controlled by the appetite. Under these conditions it is easy to expend more energy if necessary. When food is scarce, however, and energy intake limited, physical activity is reduced to more or less match the reduced energy intake. Under very severe conditions the physical activity may fall so much that even necessary tasks may be abandoned and survival thereby threatened. It is clearly not useful to ask starving people to work harder, nor somebody dieting on a very low calorie intake to do more exercise.

After a period of severe undernutrition the body may accumulate water (oedema), making body weight an inadequate guide to the degree of wasting. When such a starving person is given a good diet the accumulated fluid is excreted in the urine causing a fall in body weight, which may alarm the subject who, unless warned, expects to gain weight as soon as extra food is eaten.

Effect of pregnancy and lactation

This is described in Chapter 3.

Chapter 2

Obesity and weight control

Obesity is almost always brought about by the intake of more energy (food) than is necessary for day-to-day living. The excess energy consumed is stored as fat and this can be lost only by using up more energy than is eaten (excluding the surgical removal of fat). The fat may be laid down quite slowly or it may be accumulated rapidly. Once fat has been stored, eating just enough each day to balance energy output will leave accumulated fat unchanged. Obese people may therefore correctly state that they now eat only as much as many thin people yet they remain fat. As many obese people are inactive and because they are well heat-insulated by the fat under the skin, their food requirements may indeed be quite small but they must nevertheless eat less energy than they expend if they are to lose weight.

During the 1980s and 1990s the intake of energy in the United Kingdom fell but the fall in physical activity was even greater resulting in an increase in general body weight. Almost everybody in the United Kingdom is aware of obesity: it can be recognised easily by most people, there is repeated reference by the media and governmental offices of the need to control it, but there is only very little success in dealing with the problem.

Control of body weight

There is a general view that getting fatter on getting older is natural and it is certainly what happens with many people who can afford an ample supply of palatable food. But not everybody gains weight on ageing. In fact, some people maintain a remarkably stable weight over their entire adult life apart from fluctuations that occur during bouts of illness or during pregnancy and lactation. The way in which body weight is controlled is not understood in sufficient detail, but it is known that in the part of the brain called the hypothalamus there are two areas (centres), one of which causes eating to continue (the feeding centre) and the other causes eating to stop (the satiety centre). If the feeding centre is damaged, an animal will starve to death even though there is plenty of palatable food available. On the other hand, if the satiety centre is not working properly, an animal will continue to eat until it becomes grossly obese. Similar centres in the brains of humans work in the same way but damage to them is very rare and unlikely to be the cause of obesity.

In addition to these nervous centres, there is at least one chemical in the blood which helps the centres to control energy balance. This substance is a protein called leptin, which is secreted into the blood by the cells which store fat (adipocytes). It is encoded by the *ob* (obesity) gene. Leptin passes via the blood to the centres in the hypothalamus and signals either how much fat there is in the body or the rate of fat formation or both of these. With a rise in blood leptin there is normally reduction in eating and a simultaneous fall in insulin and cortisol secretion into the blood, which causes a fall in fat formation by a direct action on the fat cells. There is, in addition, a rise in energy expenditure. In obese people, instead of there being a lack of leptin there is an increase of it in the blood, indicating that the hypothalamus has become less sensitive to its signal, possibly because it does not pass sufficiently well from the blood to the brain cells or because the cells have become resistant to it. On very rare occasions there may be a defective *ob* gene present, causing a fall in leptin in the blood and the expected gross obesity.

Obesity is often seen to occur in family groups and it is not always easy to decide whether the cause is chiefly genetic or chiefly cultural. The discovery of the rare abnormal *ob* gene and its effect on blood leptin is evidence for some genetic basis of obesity. Another is the observation that infants separated from their biological parents at birth develop body weights more like their biological parents than like the body weights of their adoptive parents. The greatest similarity is between mothers and daughters. It has also been found that identical twins generally keep remarkably similar body weights over the years.

All the known mechanisms for weight control cannot completely account for the remarkable control of body weight that exists over many years. None of them seems likely to adjust the energy intake so carefully as to prevent a daily excess of 50 kcal being eaten, yet such an excess would cause severe obesity over time unless an as yet unknown and very delicate controlling system were present.

It has been suggested that excess food can be burned off by special cells (called brown fat) in order to keep body weight constant and that obese people may be deficient in this regard. Although this mechanism, called diet-induced thermogenesis (DIT), may operate in some animals it seems not to be of importance in humans.

What constitutes obesity?

Obesity exists when the stores of body fat are excessive. It must not be confused with a high body weight because of bulky muscles or excess body water.

The true measurement of body fat is not feasible in day-to-day investigations but an assessment can be made by use of the body mass index (BMI). To derive the body mass index the subject's weight in kg (with light indoor clothing but without shoes and with the bladder emptied) is divided by the square of the

Body mass index

Below 20: probably too thin
20–25: good
25–30: mildly overweight; usually no treatment necessary
30–40: obese; should lose weight
Over 40: gravely obese; needs urgent treatment

height measured in metres. For example, an average adult who is not obese might weigh 70 kg and have a height of 1.8 m. The height squared in this example (1.8 × 1.8) is 3.24 and dividing the weight of 70 kg by 3.24 gives a body mass index of 21.6. This is also called Quetelet's index. People with values of 20–25 are considered to be in the desirable range because they have the lowest mortality rate (from all causes). As the index rises above 25 the mortality rate also rises. For values of 25–30 the mortality rate rises only slowly and these people are considered to be only mildly overweight and probably do not need corrective treatment unless they have a wish to be thinner. If the index goes above 30, however, the mortality rate rises more steeply, so that somebody with an index of 30–35 is probably being damaged by their obesity. Those whose index is above 40 are gravely obese and need urgent medical attention.

Another assessment of body fat can be obtained by measuring the thickness of pinched-up skin over the upper arm, the hip and the shoulder-blade. This method requires a special device for measuring the thickness, some skill and is not always convenient. It is not useful when the skin is tense, as in marked obesity.

The above two methods are useful as quick convenient guides, but the results must be interpreted carefully for each subject. For example, when a non-obese but very muscular person of 80 kg with a body mass index of 27 becomes inactive much muscle may be replaced by fat. The body weight may still be 80 kg and the index remains at 27. Despite having the same body weight of 80 kg and an index of 27 throughout, this subject has passed from being not overweight to being mildly overweight as muscle is replaced by fat.

Losing weight

Many people attempt to lose weight and most are successful to some degree for a while, but after a year or so almost all are back to their original weight. Losing weight and staying thinner is clearly a difficult thing to do.

Before starting on a slimming regime the subject should be sure that there is a need to lose weight, that there is sufficient motivation to do so and that it is understood that eating habits will have to be changed permanently and that circumstances will allow for this. Further, although losing weight may bring better health, better appearance, more comfort and improved job prospects, it will not necessarily bring happiness or cure all problems.

At the start of a slimming regime a realistic target should be set and some degree of flexibility allowed for. Some people lose weight more easily than others. For most, there is no urgency to lose weight and a reduction of about 0.5 kg

Losing weight

In mild overweight, losing about 250 g (about 0.5 lb) per week is a satisfactory rate: losing more than about 500 g (about 1 lb) per week is often too much.

(1 lb) per week after the first four weeks can usually be achieved without causing much hunger or loss of muscle power. During the first week or two there is a loss of stored carbohydrate (glycogen) and its accompanying water, amounting to about 4 kg, after which the fall in weight slows down. This has nothing to do with the amount of water taken. No attempt should be made to lose weight by drinking less fluid, which can be harmful and has no effect on the rate of loss of body fat. Diuretics should never be taken to reduce body weight.

Losing weight at a slow rate has two advantages. First, the reduction in food intake is small and most people can soon accommodate to it. Second, it enables the subject to get used to a new style of eating over a substantial period and this new eating pattern can gradually and permanently replace the previous one.

Body weight should be measured about once a week, using the same scales in the morning before eating or drinking and after emptying the bladder. Change in body water and in the weight of the contents of the gastro-intestinal tract can together cause a 0.5–1.0 kg variation each day even on a constant food intake. It may take a few weeks to find the amount of food needed for the rate of weight loss desired.

The diet should be as varied as possible with only a little fried foods, fatty meat, full-fat cheese, biscuits, cakes, fat spreads, snack food and alcohol. Intake of fruit and vegetables should be increased.

Diets designed to achieve rapid loss of weight can be dangerous and do not teach a satisfactory eating habit. Unlike the very small effect that losing weight slowly has on general well-being, the debilitating effect of rapid weight loss can be very marked. People taking only about 1800 kcal/day, which is near the resting metabolic rate, have a fall in metabolic rate, a fall in the pulse rate, a decrease in general activity and a decrease in tolerance to cold. All these changes are attempts by the body to conserve energy and result in a diminution of the rate of weight loss. In addition, a much reduced energy intake may produce a constant anxiety about food, irritability, lack of interest in everyday things and some degree of depression. These mental changes may persist for months after normal eating is resumed.

If, during dieting, much exercise is taken, muscles may hypertrophy, so that although fat is being lost, body weight may not be declining very much. An inactive dieter may show greater weight loss but the very active person will probably be much fitter.

People only mildly overweight (body mass index 25–27) may find that the small advantages of their losing weight are not worth the effort of dieting and will very likely be better off doing nothing. This is especially so for the elderly.

A body mass index greater than 35 needs specialist advice immediately at an obesity clinic. In addition to being obese there are very likely medical problems requiring attention.

There are no such things as slimming foods: all foods will result in overweight if enough is eaten. No food can make you thinner. Pills and preparations to lose weight should never be taken except on medical advice.

Complications of obesity

Obesity can cause both physical and psychological damage and is associated with a decreased life expectancy. Of the numerous medical conditions found in obesity, diabetes mellitus is perhaps the most important. It is about five times more likely to be the cause of death in obese men than in thin men and it causes death nearly ten times more often in obese women than in thin ones. In addition, before these diabetics die they often have several years of poor quality life brought about by the diabetes. Losing weight, especially in the earlier stages of the disease, brings improvement and, even if this is not marked, further complications may be avoided for many years.

Also found in obese people, particularly younger ones, is ischaemic heart disease which may be accompanied by high blood pressure (hypertension), although this latter condition is not itself now thought to be due directly to obesity. Losing weight brings improvement in the ability to climb stairs, carry parcels, walk uphill and to occasionally run. Part of this improvement in exercise tolerance on losing weight is due to the respiratory system becoming healthier.

An important problem common in obesity, although not life-threatening, is damage to weight-bearing joints. Osteoarthritis of the hips, knees and ankles along with damage to the feet limits mobility and causes pain.

Gall-bladder disease is more prevalent in overweight people and its treatment less satisfactory.

Surgical operations are often more hazardous and the outcome likely to be poorer in the obese because operations may be more difficult for the surgeon, while the anaesthetist may have to cope with a failing heart and an inadequate respiratory system.

Other conditions commonly found in obese people are varicose veins, stretched skin causing permanent disfigurement, and irritation and infection of skin produced by chronic accumulation of sweat between folds of skin and under the breasts.

As so many of the problems associated with obesity limit mobility, sometimes severely, the exercise option for aiding weight loss is often not very useful, which means that reduction in food intake must be greater than it would be in more mobile subjects. Markedly obese people are usually very inactive, even fidgeting being almost absent. This physical inactivity plus their better heat insulation results in the obese needing relatively little food. However, if an obese person is able to be active, the physical effort of carrying the excess fat uses up much energy and thereby greatly aids weight loss.

In addition to these ailments, obese people quite often have emotional problems because they find it more difficult than thin people to find a spouse, get a job, partake in sports and sometimes in travelling by public transport or even by private car. In western societies there is often considerable prejudice against the obese.

Natural tendency to overweight

It is not generally possible to foretell among the young who are going to become overweight. The fate of frank over-eaters is often clear, but over-eating need be only small for it to culminate in overweight. Inadequate energy expenditure for the amount of food eaten may also be small and not at all obvious before overweight becomes apparent. Many people are able to match their energy intake and output so as to remain almost the same weight over many years but some seem to lack this ability. Whether overweight is mainly an inherited trait is not known but it is possible. Becoming moderately overweight has survival value, enabling energy to be stored when food is plentiful as a precaution against starvation when food is scarce. It is only in affluent societies that moderate overweight becomes unnecessary.

Anorectic drugs

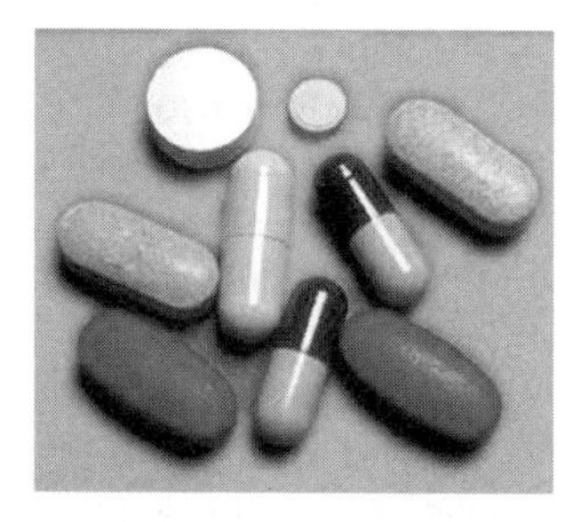

These are substances that help in weight loss by decreasing appetite. They are mainly amphetamine and its derivatives and all have undesirable side-effects. Experience with them is that, although they may help to control hunger and increase weight loss at first, they generally lose their efficacy after a time. So far, there is no drug which will produce effortless weight loss without side-effects. The current drugs may help some individuals in the initial stages of dieting but the only satisfactory way to produce a permanent reduction in body weight is for the subject to learn to eat a good mixed diet in the quantity necessary to maintain a body mass index of 20–25. Anorectic drugs should never be taken except under the guidance of a medical specialist.

Metabolism-boosting drugs

The most efficient of these substances are the thyroid hormones. However, doses which increase metabolism sufficiently to produce a useful loss of weight usually cause undesirable side-effects, particularly damage to the heart. Furthermore, they often cause an unwanted loss of the non-fat tissues of the body rather than just a loss of fat. The only time when thyroid hormones are needed by obese people is when they have a poorly functioning thyroid, but this is uncommon.

Slimming groups

These are of particular value to people who are only moderately obese (body mass index 25–30). Some groups are run for profit while others are on a non-profit basis. The techniques used by different groups vary a lot, so it is important to find a suitable one. The main advantage of joining a slimming group is the mutual support provided by its members. The exchange of ideas and advice helps members to think constructively about their problem and the feeling of not being alone helps some people. The need to eat less and exercise more still remains.

Childhood obesity

It can be difficult to determine minor degrees of obesity in childhood although grosser states can be defined. It is sometimes said that fat children become fat adults but it does not seem to be true. Of a group of obese 36 year-olds only about one-quarter were obese as children. This suggests that treating moderately obese children would not drastically reduce the number of obese adults. As many children change their degree of obesity several times over a few years, minor overweight should be ignored in otherwise healthy children.

Much childhood obesity seems to be familial or cultural and one of the ways to help limit overweight in children is to prevent sweets, cakes and biscuits being used as bribes or as a solace. If a child is taught that food can compensate for unhappiness, adult obesity may well be the result.

If it is thought necessary to limit the rate of weight gain in a child, expert medical advice should be sought. Unlike an adult, where simply keeping an unvarying weight is the goal, a child must be allowed to gain weight and designing a restricted but healthy diet for this is difficult. Encouraging a child to take plenty of exercise is better than merely limiting the energy intake and it may help the child to take a keen interest in its physical abilities and appearance. This is more likely to have long-term success than simply restricting the diet.

Dieting in pregnancy

This is dealt with in the following Chapter.

Chapter 3

Pregnancy and lactation

The general principles of nutrition apply to both the pregnant and non-pregnant woman and most of the Chapters of this book are therefore of importance to all pregnant women and those likely to become pregnant. This Chapter deals with some of the special aspects of nutrition during pregnancy and lactation.

The health of a pregnant woman and her developing baby depends greatly on the quality of her pre-pregnancy diet as well as her diet throughout the pregnancy. In addition, a good diet is essential after delivery of the baby to enable the mother to breast-feed for 5–6 months. A deficient diet may lead to malnutrition of the mother and to a small baby at greater risk in the early months of life. During and following the Second World War there were several epidemics of malformed children in various parts of Europe: these epidemics were attributed to poor nutrition before and during pregnancy. For women severely malnourished pregnancy is very likely to end in miscarriage.

Women who have little body fat, which may be due to inadequate energy intake or to extreme exercise, or a combination of these factors, stop menstruating (amenorrhoea). They may become infertile even before this stage is reached. The amount of body fat governs female fertility and may have a survival value, because if a grossly undernourished woman does become pregnant she is likely to have an abnormal baby and her own life may also be in danger. For a woman to have regular menstruation she needs about 22 per cent body fat and to be normally fertile she needs about 25 per cent. Thus an 18 year-old woman of 165 cm height (5 ft 5 in) and who is not unusually muscular needs to weigh about 49 kg (108 lb) for menstruation to begin and she would need to weigh about 51 kg (112 lb) to be normally fertile. Taller women would need to weigh more and shorter ones less.

Energy cost of pregnancy

During pregnancy there is a gradually increasing need for energy to provide for the growth of the developing baby, the uterus, the placenta and the enlarging breasts. Near the end of pregnancy the breasts start to secrete a very rich special secretion (colostrum) which precedes the production of mature milk, further increasing the need for energy. In addition to these calls for extra energy, the pregnant woman normally accumulates about 4 kg (9 lb) of fat, which should be used up later during breast-feeding if that is continued for 5–6 months. All this extra energy requirement plus an allowance for greater physical effort during pregnancy may amount to as much as 80,000 kcal being needed during the 40 weeks of pregnancy. A pregnant teenager who is herself still growing may need even more. Many women limit the accumulation of extra fat, especially if they do not intend to breast-feed the baby for more than a few weeks and they often also limit the extra physical effort by obtaining domestic help so that only an extra 40,000 kcal may be required over the forty weeks.

The extra food needed during pregnancy is quite small during the first three months, larger during the middle three months (an extra 100 kcal/day) and greatest during the last three months (an extra 200 kcal/day). It is probably preferable to smooth out this change in requirement, after any period of morning sickness, by eating more than is necessary in the early to middle period of pregnancy because in the later weeks the increased size of the uterus and its contents often make the eating of large meals difficult. Total body weight gain should be steady and a typical rate is 2 kg (4.5 lb) in the first three months and then about 0.25 kg (0.5 lb) per week for the rest of pregnancy, making a total of 9–10 kg (20–22 lb). Care must be taken to prevent excessive weight gain especially towards the end of pregnancy because it is associated with pregnancy hypertension (high blood pressure) as well as post-pregnancy obesity.

Despite all the calculations about how much pregnant women should eat to cover the energy costs of pregnancy, some women, especially if well-nourished before pregnancy, seem to eat little extra and yet have healthy full-term babies.

The food intake, increased during pregnancy, must be reduced to the pre-pregnancy level soon after the birth of the baby and most of the extra energy needed for breast-feeding should come from the stored extra fat. In this way the mother should return to her pre-pregnancy weight by the time the baby is weaned.

Energy cost of lactation

Lactation has about twice the energy cost of pregnancy, the latter needing about 300 kcal/day while 600–700 kcal are needed each day for lactation. The variation between women is wide, some producing as much as one litre or more milk per day, while others produce only half a litre. The average is about 800 ml/day, providing about 600 kcal each day for the baby, although this value also varies considerably. The calorie value of breast milk and also its nutrient contents may vary appreciably from day to day for any individual mother. Because milk production is probably only about 80 per cent efficient the mother has to use about 750 kcal to produce the 800 ml of milk. Not all the energy needed for lactation should come from extra food, especially if the mother has gained a few kilograms of extra fat during the pregnancy. This extra fat should be used to provide energy for milk production if the mother is to return to her pre-pregnancy weight. Thus if the lactational energy need each day is about 750 kcal, the mother should eat about 500 kcal extra each day and use up about 250 kcal each day from her fat reserves. In this way her weight will gradually fall and she will not become fatter with each pregnancy. Looking after a newborn child will also help weight loss.

Mothers who do not intend to breast-feed their baby should take care to limit their fat gain during pregnancy because they may find it very hard to lose weight at the same time as they have to care for a newborn child.

The best way to measure how much milk a baby has had is to weigh it together with its clothes before and after feeding and the difference in the two weighings gives the amount of milk consumed. The napkin should not be removed until after the second weighing even if it becomes soiled. The only way to judge if the milk volume and quality are adequate is to make sure that the baby is gaining weight normally and is judged by an experienced person to be developing satisfactorily.

During lactation the nutritional quality of the milk is kept more or less constant by using the mother's reserves if necessary. Hence a satisfactory diet before, during and after pregnancy is essential to prevent lactation from seriously depleting the mother's tissues. This applies to almost all the major dietary items except iron, which is always low in breast milk. After the birth of the baby there is little further drain on the mother's iron reserves, although supplements are sometimes continued for a few months.

Dieting in pregnancy

> Pregnancy and lactation need a good supply of protein, iron, calcium and vitamins, especially folate.
> Liver, peanuts, soft cheeses and pâtés should be avoided.

The natural course of events in pregnancy is for the healthy mother to gain about 9–10 kg in body weight. This is brought about by growth of the breasts, by growth of the uterus and its contents, by an expanded blood volume and by the accumulation of about 4 kg of body fat. This gain requires extra food intake, with emphasis on an increase in high-quality protein, iron, calcium and vitamins. The extra 4 kg of fat will be needed to support lactation for about 5 months after the birth of the baby, during which time the mother should be gradually reverting to her pre-pregnancy food intake. Under these conditions, at about six months after the birth the mother should be about the same weight as she was at the start of her pregnancy. If the mother does not breast-feed her child for about 5 months then some or all of the extra 4 kg of fat she accumulated during pregnancy will have to be lost by her eating less if return to normal weight is to occur. If the mother is convinced that she is not going to breast-feed her child for several months she needs to limit her weight gain during pregnancy. This needs expert advice so that no harm is done to the developing baby. Babies of well-fed mothers are larger and healthier than the babies of under-fed or badly-fed mothers.

It may be necessary to consider dieting in pregnancy for women who are sufficiently obese to cause anxiety about complications during pregnancy which are more common and more hazardous in obese women. This should be undertaken only after careful assessment at an obesity clinic. Under such supervision the health of the mother during and after her pregnancy can be improved and no adverse effects will ensue.

Calcium and vitamin D

Although as much as two-thirds of the baby's calcium is laid down during the last three months of pregnancy, the mother should be sure to have a plentiful calcium intake throughout the pregnancy. The surplus calcium will be stored in the mother's bones and will be used later in the pregnancy and also during lactation. The diet should contain 1200–1500 mg calcium per day during pregnancy and this should be increased to 1500–2000 mg during lactation when the need for calcium rises considerably. In the non-pregnant woman about 30 per cent of this intake would be absorbed into the blood but this is increased to about 40 per cent or even more during pregnancy and lactation. Because efficient absorption of calcium by the intestine requires an adequate supply of vitamin D a daily supplement of 10 μg (400 i.u.) of the vitamin should be taken throughout pregnancy and lactation. A diet rich in calcium and vitamin D will keep calcium depletion of the mother's bones to a minimum and will thereby help reduce later osteoporosis.

There are 250–350 mg of calcium in a daily output of about 800 ml of breast milk. If the absorption of calcium during lactation is 40 per cent, there is a need for a daily dietary supply of about 800 mg of calcium, which is contained in about 700 ml (1.2 pint) of cows' milk or in 115 g (4 oz) of hard cheese. This calcium requirement is, of course, in addition to the mother's other needs and hence her total intake of calcium should be in the range of 1500–2000 mg. Cows' milk is very rich in calcium and has about 120 mg per 100 ml, in contrast to about 35 mg in 100 ml of breast milk. Unless the mother is badly in need of extra energy, she can take skimmed or semi-skimmed cows' milk instead of full-fat milk so as to limit her intake of fat.

If for some reason it is not possible to take cows' milk or hard cheese each day, then calcium supplements in tablet form can be used.

Pregnant and lactating women should avoid foods high in phytate because this can reduce the absorption of calcium from the intestine.

Iron

Over the whole of pregnancy there is a need for about 900 mg of iron, which is about 3 mg of iron each day for the 280 days. To absorb about 3 mg of iron from the intestine an intake of at least 30 mg of iron is needed because iron is absorbed relatively poorly, even though it is higher in pregnant women than in those not pregnant. In general, it is better to obtain the iron from an iron-rich diet, usually based on meat or iron-fortified foods. If such diets are not feasible then iron tablets should be taken, although these sometimes cause abdominal discomfort. Some authorities recommend routine iron supplements as soon as pregnancy begins because most ordinary diets do not supply sufficient iron and many women have low iron reserves. During the second half of pregnancy, when iron needs are greatest, many women show little stores remaining in their bone marrow. For this reason women whose diet is not rich in iron or who show signs of being iron-depleted should, under supervision, take a daily tablet of ferrous sulphate or similar compound to augment their iron intake.

Any woman who is likely to become pregnant should have a regular iron-rich diet so that she starts her pregnancy with good iron reserves. When devising an iron-rich diet for pregnancy it is important to omit the use of liver because liver may contain excessive amounts of vitamin A, which can seriously damage the developing baby. It is worthwhile continuing with an iron-rich diet for at least three months after the baby's birth to ensure that the mother's stores are replenished. Iron tablets should not be taken regularly without professional guidance because excessive iron intake in this form can be harmful.

Breast milk contains little iron and lactation does not exert a drain on the mother's reserves. As there is often no menstruation during lactation the mother can usually restore her iron levels on a good diet without the need for supplementation. During pregnancy the mother's blood contains more red blood cells (which hold the iron-containing haemoglobin) than when non-pregnant. After birth, the unneeded red cells are broken-down and the iron in them re-used, acting as a supplement to the iron in the diet.

Foods high in phytate can reduce the intestinal absorption of iron.

Folate (folic acid; folacin)

During pregnancy an adequate supply of folate to the developing baby is essential for the proper growth of the nervous system; without it, conditions known as neural tube defects occur. As the neural tube is formed within the first four weeks of pregnancy, the damage that occurs with an inadequate folate intake may well take place before the mother knows she is pregnant. This situation is aggravated by the low level of folate in many non-pregnant women, although it is unusual for the level to be so low as to produce a megaloblastic anaemia, which would alert them to the deficiency. For these reasons all women who may become pregnant should eat a diet rich in folate. If they become pregnant they should supplement their diet with a daily tablet of 0.4 mg of folate for at least the first twelve weeks of pregnancy. Women who are trying to become pregnant should start the daily folate supplementation before they conceive. For women who have already had a baby with a neural tube defect the daily folate supplementation is increased ten-fold to 4 mg per day. Supplementation does not prevent neural tube defects entirely but it does diminish the incidence by about 70 per cent. As folate is readily excreted in the urine there is little danger of these doses being toxic.

Although liver is a good source of folate it should not be eaten just prior to or during pregnancy because sometimes it contains very high levels of vitamin A which may damage the developing baby.

The recommended daily dietary folate intake during pregnancy is about 0.8 mg and during lactation about 0.6 mg. Many popular diets do not reach these levels. Alcohol interferes with folate absorption and utilization and regular alcohol users should increase the folate eaten. Some medicines also increase folate need.

Vitamin B_{12}

Dietary deficiency of vitamin B_{12} is extremely rare except in strict vegans because although the vitamin does not occur in plant foods it is present in virtually all foods of animal origin. The extra amount needed for pregnancy and lactation is unknown because in even a poorly nourished but healthy non-vegan mother her liver almost always has a store of vitamin B_{12} which will last for several years and only a small amount will be used up during the pregnancy and lactation. Nevertheless, some authorities quote the daily dietary requirement during pregnancy as being 3–4 μg (micrograms), of which about half will be absorbed. A mother drinking 600 ml (1 pint) of cows' milk each day will absorb 1–2 μg from that source alone.

Unless a mother is B_{12}-deficient, the amount of the vitamin in her breast milk will provide 0.2–0.8 μg per day, which is ample, but an extra 2.5–4 μg per day has been recommended. For formula-fed infants the recommended intake is 0.15 μg per 100 kcal, so that a baby getting about 800 ml milk per day would have about 650 kcal per day, making the B_{12} requirement about 1.0 μg per day, which is very generous. If the mother is B_{12}-deficient a supplement of 0.3 μg B_{12} is recommended for the baby from birth to 1 year, although 0.1 μg would probably suffice.

Although liver is an especially rich source of vitamin B_{12} it should not be eaten by women likely to become pregnant as it may contain excessive amounts of vitamin A, which can harm the developing baby.

Iodine

The recommended daily intake of iodine by the mother during pregnancy and lactation is 175–200 μg per day. This is a very safe amount as intakes up to 1000 μg per day appear harmless. The normal UK diet provides ample iodine, especially if some salt-water fish is eaten two or three times each week. Meat, eggs and most vegetables are also useful sources and iodized salt should be used for cooking and at the table. Iodine lack was more common when all the food came from one area which might have an iodine-deficient soil, but now that foods from many different areas are regularly eaten the incidence of iodine-deficiency in the United Kingdom is very low.

If a mother does become iodine-deficient, not only may she show abnormalities but her baby may be born with hypothyroidism (cretinism). Such infants are dwarfed, have a thick dry skin, a large protruding tongue and are mentally retarded. With immediate treatment the infant may become normal, but delayed treatment will result in a mentally and physically retarded child.

Alcohol

Alcohol taken during pregnancy, except in very small amounts, may cause damage to the developing baby. This may lead to the foetal alcohol syndrome, in which the baby is born with a small deformed head, is underweight and is mentally retarded due to alcoholic damage to the developing brain. Although there is a view that alcohol in moderation during pregnancy may be harmless, the wisest behaviour for a pregnant woman is to avoid alcohol. Binge drinking is particularly harmful.

Protein

All the numerous proteins of a new baby are normally made from components of the mother's diet. During the first few months of pregnancy the essential and the non-essential amino acids for making the proteins have to be supplied by the mother, but after about 20 weeks the developing baby is able to synthesize the non-essential amino acids for itself, although it still needs a supply of the essential amino acids. Mothers on a good high-protein diet have more successful pregnancies than those on a low-protein diet. If the dietary protein is inadequate, protein is transferred from the mother's tissues.

The total protein of a newborn baby weighs about 0.5 kg. This plus the protein needed for the placenta, the enlarging uterus and the breasts amounts to about 1 kg. In addition, there is the protein needed for extra plasma proteins and extra red blood cells in the mother's expanded blood volume. The mother's diet therefore needs more protein than she might normally eat when non-pregnant and she should increase her normal protein intake by about an extra 30 g per day. This is done by the use of 600 ml (1 pint) of skimmed or semi-skimmed milk (to keep the daily fat intake down), giving about 20 g of protein, plus an extra 100 g (3.5 oz pre-cooked weight) of meat, poultry or fish, giving a further 18 g of protein each day throughout the pregnancy. For vegetarians a diet rich in cereals and pulses must be used, with milk, cheese and eggs if taken. If the pregnant mother is an adolescent a further 15–20 g of protein are necessary each day to allow for the needs of the still growing mother.

The protein of breast milk is of very high quality and has an essential amino acid to non-essential amino acid ratio of almost 1:1, which the infant needs. This is in contrast to a ratio of about 1:4 in a satisfactory diet for an adult. Breast milk protein needs no protein supplementation for the infant's first six months. During lactation the mother's protein intake should be increased by about 20 g per day.

After weaning, an infant's diet should be protein-rich by feeding meat, poultry, fish, eggs, cheese, cereals and pulses as well as formula milk. Although cows' milk is a good source of protein it should not be used before the child is 1 year old.

Heating protein may diminish its nutritional quality or it may improve it. If heated in the presence of some reducing sugars a reaction occurs which makes the essential amino acid lysine unavailable and may sometimes do the same for the amino acids arginine, tryptophan and histidine. However, this does not occur with the reducing sugar lactose found in milk and the boiling of milk, essential for feeding to infants, actually makes the milk protein casein more easily digestible than is the case in unboiled milk. Toasting cereals reduces the protein quality but if it makes the food more palatable it may be advantageous. Repeatedly cooking protein may destroy the essential amino acid methionine.

The damage done to an infant by a diet inadequate in protein is dealt with in Chapter 18.

Vitamin A

The normal daily UK recommended intake of vitamin A (or its equivalents) for adults and pregnant women is 750 μg (2500 i.u.) and this is increased to 1200 μg (3600 i.u.) for lactation. If pre-formed vitamin A (retinol) is being taken the maximum daily intake during pregnancy must not exceed 3000 μg (9000 i.u.) because intakes above this may damage the developing baby, especially during the first nine weeks of pregnancy, producing abnormalities of the face and head. If vitamin A activity is being obtained from only β-carotene and other plant carotenoids there is no danger of damage to the baby.

Because of the problem of vitamin A overdosage, pregnant women and those likely to become pregnant should not eat liver or foods containing liver (sausages, pâtés, pies) as they may contain very large amounts of vitamin A. Some samples of liver have had as much as 30,000 μg per 100 g of liver, so that a serving of 100 g (3.5 oz) would supply ten times the maximum vitamin A allowance. In addition, vitamin A supplements must not be used except when under specialist supervision. A single accidental overdose of 150,000 μg in the second month of pregnancy has resulted in a deformed infant.

Tretinoin and isotetrinoin, used for acne treatment, are similar to vitamin A and must be avoided just before and during pregnancy.

The danger of vitamin A overdose during pregnancy arises only when preformed vitamin A (retinol) itself is used, whereas there seems to be no danger with β-carotene. Pregnant women should therefore eat a liberal helping of well-cooked carrots three to four times a week and also have some dark-green or yellow-red vegetables each day. In this way they will obtain all the vitamin A activity they need from carotenoids with no fear of overdosing.

Vitamin C

Breast milk is relatively rich in vitamin C, containing about 6 mg of the vitamin in 100 ml, so that on average an infant fed only breast milk gets about 50 mg of vitamin C per day. To ensure this supply the mother's vitamin C intake during pregnancy should be about 100 mg per day and that should be increased to about 150 mg per day during lactation. This can be obtained from three to four glasses of good quality orange juice per day or from vitamin C tablets. Orange juice is sufficiently acidic to damage the teeth unless they are rinsed soon after drinking the juice. Large overdosing with vitamin C by a pregnant woman may lead to the developing baby becoming tolerant to the vitamin and it will then require more than the normal amount for a while after birth, with the danger that scurvy may develop on only the usual amount of vitamin C in the breast milk.

A newborn baby of a well-fed mother has a vitamin C store which is enough to prevent scurvy for about five months provided the baby has few infections.

If the baby is bottle-fed, because some vitamin C will be destroyed on boiling the milk, infant formula milk well-fortified with vitamin C should be used or vitamin C supplements added to the milk.

Infantile scurvy is rare in the United Kingdom and is unlikely to occur unless the mother is badly malnourished and the infant weaned early on an inadequate diet. The most marked damage is deranged bone growth at the ends of the long bones and ribs, with painful swelling around the joints, especially in the arms. The infant tends to lie mainly on its back and cries if handled (a normal crying baby tends to stop crying on being handled). There may be bleeding into the skin, stools and urine. The condition rapidly improves after a few days on vitamin C supplements.

Vitamin B_6

The recommended daily intake of about 2 mg of vitamin B_6 for normal adults is raised to about 2.5 mg per day for pregnant women and sometimes to 5 mg per day during lactation. Breast milk is relatively poor in vitamin B_6 and unless the mother's intake of the vitamin is increased her milk may not provide the daily 0.3 mg vitamin B_6 needed by the infant before weaning. An ordinary UK mixed diet almost always needs some supplementation during pregnancy and lactation. The supplementation should not be greater than 10 mg vitamin B_6 per day because higher levels may cause nerve damage.

Vitamin B_6 has been given for nausea in pregnancy but it is not often effective.

Listeriosis

Listeriosis is an infection caused by eating food contaminated with the listeria organism. The disease produces symptoms like those of influenza and in healthy adults is usually mild. In pregnant women, however, the organism can gravely infect the developing baby and may cause miscarriage. If infection occurs just at the end of the pregnancy the newborn baby may be severely ill and is likely to die. The foods which are most often contaminated with listeria are soft cheeses (such as brie, Camembert and blue-veined varieties) and all types of pâtés. The listeria organism is unusual in that it can multiply at low temperatures so that keeping food in a refrigerator does not help once the food has been contaminated. It is safe and advisable to eat the hard cheeses (such as Cheddar) because the listeria organisms do not survive in these. Processed cheeses, cheese spreads and cottage cheese are also safe. Ready-to-eat foods should be re-heated until they are very hot, especially in the centre, before being eaten by a pregnant woman. Heating to above 70°C for two minutes will kill listeria.

Listeriosis can also be contracted by contact with sheep at lambing time. It may also be transmitted by contaminated silage and its products.

Medicines

Many medicinal products are excreted in breast milk and many may have a harmful effect on the baby. Self-medication should therefore be avoided and if the mother is prescribed a medicine by her doctor she should emphasize the fact that she is breast-feeding her baby, so as to alert her doctor to the possible need to change to bottle-feeding.

Peanuts (Groundnuts)

Pregnant and lactating women who have members of the family who suffer from hay fever, eczema or other allergic responses should not eat foods containing peanuts. If no family members have these conditions peanuts may be eaten.

This advice is given in order to help reduce the rising numbers of children who have peanut allergy. Currently about one person in two hundred in the United Kingdom has this allergy and about six die each year as a result. It seems that the developing baby can become sensitized to peanut antigens in the mother's blood. Whether this recommendation will reduce the number of adults with peanut allergy will take some years to decide.

Chapter 4

Infancy (0–1 year of age)

The best food for a healthy baby of a healthy mother is breast milk. It is made for the job, is at the right temperature, is safe and is almost free. If the mother can produce enough good quality milk it is all that is required for the first four months of life.

For the first few days after the infant's birth the breast milk is a yellowy colour and is almost clear; it is called colostrum and is very rich in proteins needed for growth and to combat infections. The milk gradually becomes thinner and bluey-white; this is mature milk. After two to four weeks the breasts produce their maximum output, around 600–900 ml (1–1.5 pints) per 24 hours. Many mothers find that this naturally increases their thirst so that they drink 1–3 litres (2–4 pints) of fluid each day. At least 0.5 litre (1 pint) of this should be cows' milk (low-fat if the mother needs to control her weight), which provides calcium as well as other essential nutrients. Suckling (sucking at the nipple) is itself a cause of further secretion of milk so that two-hourly feeding is desirable during the first few weeks. After a few weeks the infant develops its own routine of feeding depending mainly on how much milk it gets at each feed and this timing is better than a routine imposed upon the infant by a clock. For the first two months most infants need feeding every two to four hours, but after that the night feed can often be omitted. Infants vary a lot in their feeding habits and experience with a particular infant is the best guide. Even if breast feeding is not going to be continued for long, a few weeks at the beginning of an infant's life is better than nothing.

> Cows' milk should not be used as a main drink before 1 year of age.

If breast milk is not available then infant formula milk must be used for the first six months of life. Thereafter, infant formula milk may be continued or follow-on formula milk may be used. Ordinary cows' milk should not be used as a main drink under one year of age. After six months of age boiled unmodified cows' milk may be used in making food such as custard and small amounts can be used for occasional drinking and on breakfast cereals. When bottle feeding it is essential that there is an adequate supply of good quality infant formula milk, that the instructions are followed accurately, that all utensils are properly sterilized and that there is a good supply of pure water.

Except for breast milk, all other milks must be pasteurized or boiled. When making up formula milk it is essential to read the instructions very carefully. Only the correct amount of powder should be used because a too concentrated milk solution may be very harmful to the infant. In addition, the excess energy fed by using too much powder is likely to make the infant too fat. Formula-fed babies are more likely to be overweight than breast-fed babies.

The amount of breast milk that a baby has taken can be measured by weighing the baby (with its clothes on) before feeding and then weighing it again after feeding (still with its clothes on). The gain in weight gives the amount of milk taken; 1 g is more or less equal to 1 ml (28 g is equal to 1 oz). It is necessary to leave a napkin (diaper) on the baby all the time so that any urine or stool passed during feeding is also weighed. Measuring the milk taken at each feed is necessary only if there is doubt about the volume of milk being produced, perhaps because the baby seems to be hungry very frequently or is not gaining weight at the expected rate. It may be useful early in breast feeding to give the mother confidence that the baby is getting enough milk.

Average daily feeding

An infant needs on average about 150–160 ml milk per kg of body weight daily and this gives more or less the same energy for breast milk, modified cows' milk or infant formula milk (70–75 kcal/100 ml milk). There is, however, a considerable variation in energy need from infant to infant and satisfactory growth and development is the only sensible test of adequate nutritional intake. A placid infant may need only half the energy needed by an active infant, especially one that cries a lot.

A well-fed healthy infant doubles its birth weight in six months and trebles it in one year.

Water

The kidneys of the newborn infant are immature and take about eight weeks to develop fully their ability to excrete any excess water in the body. During this period the water in breast milk or in properly made-up infant formula milk suffices, unless the climate is unusually warm, in which case small amounts of cooled boiled water should be given. After about eight weeks a healthy infant needs about 150 ml of water per kilogram of body weight, most of which will be present in the milk fed. Boiled water may be given if the infant is thirsty. If there is any diarrhoea or if the climate is very warm extra boiled water must be given. Infants can easily become dehydrated and great care must be taken to avoid this.

Protein

Because of the infant's very high growth rate there is a need for food rich in protein, such as breast milk or infant formula milk. Between birth and three months of age an infant needs about 2.5 g of protein/kg of body weight per day, which amounts to about 12 g of total protein per day. As the infant gets older its rate of growth slackens so that between nine months and one year the protein requirement has fallen to about 1.5 g of protein/kg body weight per day, which is now about 15 g of total protein per day. Thus an infant at one year, weighing perhaps 11 kg, needs about 15 g of protein per day while its father, who weighs perhaps six times as much can keep in good health on only 50 g of protein per day, which is only about three times that needed by the baby. This is why, when foods rich in protein are scarce, the young children should be given priority over the adult men and non-pregnant non-lactating adult women.

Weaning

In the first few months of life infants can swallow only liquids and they will normally reject solid food by reflex. If any solid food is swallowed the carbohydrate and protein will not be properly digested and some of the protein may be absorbed into the blood intact, inducing a life-long allergy. Hence, although some infants do well with some solid food at ten weeks or even earlier, the UK government's committee on medical aspects of food policy (COMA) is of the opinion that the majority of infants should not be given solid food before the age of four months. Until this age, breast milk provides adequate nutrition if enough of it is produced by a well-fed healthy mother. In the absence of breast milk a good infant formula milk may be used. By about six months, however, neither breast milk nor infant formula milk is likely to provide enough protein, energy, iron, zinc and vitamins A, C and D without overloading the infant with water, so solid food has to be added to the diet. It is necessary, of course, to ensure that the infant at this stage has developed sufficiently to eat solid food.

The first solid foods should be given in quite small amounts on a spoon so that the infant can learn to accept new food. If the food is rejected no attempt to force acceptance should be tried, but the food can be given again in a day or two. Infants will eventually eat most mildly flavoured foods, especially if they see an adult apparently eating the same food. A great deal of an infant's behaviour is mimicry. Solid and semi-solid food should not be added to milk or any other fluid.

Breast or bottle feeding continued during the weaning period should be gradually diminished as the amount and variety of solid foods in the diet are increased. After weaning is completed, milk should continue to be a very important part of the diet.

Feeding from a bottle is more likely to cause damage to erupted teeth than is the use of a cup and once teeth have appeared bottles should no longer be used.

Table sugar (sucrose) should be used in only very small amounts, if at all. If used at an early age the infant may acquire an enduring liking for sweet foods, which may damage the teeth, especially the first set, and may later be a cause of childhood obesity.

Infants should always be supervised by an adult during feeding in case an emergency should arise.

Weaning diets

Commercially made weaning foods are very popular and convenient. Good brands are very safe. They are, however, comparatively expensive and are perhaps best used occasionally or when the hygienic preparation of food is difficult. Home-made weaning food can often be prepared using items of the adult diet. It is essential that very high hygienic standards are used when preparing food for infants. If there is any doubt about the safety of any food it should be discarded. Heating doubtful food does not always make it safe: the contaminating organisms may be killed but the toxins (poisons) they may have produced may remain intact and still dangerous despite the heating.

Formula milks

Formula milks are available for feeding infants when there is inadequate breast milk. These milks contain all the major constituents known to be present in breast milk and generally at the same concentration, except for increased amounts of iron, zinc and vitamin D. Other modifications may also be present.

During the first six months of life infant formula milk is used, while after six months follow-on formula milk takes its place. These follow-on milks, made of modified cows' milk, contain more protein, iron and vitamin D than does infant formula milk and are unsuitable for very young infants.

All infants, even those with plentiful iron stores at birth, are liable to develop iron-deficiency anaemia from about six months unless they are fed iron-rich foods such as meat, poultry, fish, liver, egg, beans, lentils, apricots, wholemeal bread, iron-fortified breakfast cereal and iron-enriched follow-on milk. Fresh fruit juice or a vitamin C supplement is useful with these foods to aid iron absorption. A haemoglobin estimation at around nine months is an easy investigation and will alert the mother to improve the diet if necessary.

Formula milks of infant and follow-on type are generally based on cows' milk and these should be used unless there is a special reason for not using cows' milk, in which case formula milk made from soya can be tried. The usual reason for changing to soya milk from cows' milk is allergy to protein in the cows' milk. However, because such infants can also develop an allergy to soya protein it is better to use a formula milk in which the protein has been hydrolysed (digested).

The carbohydrate in breast milk and in cows' milk is a sugar called lactose and it is more or less harmless to teeth. In formula milk, especially of the soya

Table 2 Comparison of breast milk and cows' milk

	Breast milk g/100 ml	*Cows' milk g/100 ml*
Water	87	87
Protein	1	3
Lactose	7	5
Fat (total)	4.5	4
Essential fatty acid	0.4	0.1
Calcium	0.035	0.140
Phosphorus	0.015	0.100

Breast milk and cows' milk both provide about 70 kcal/100 ml

type, other sugars may be used and these may damage teeth. After one year of age it is better, therefore, to feed formula milks from a cup rather than from a bottle. Soya-based formula milk is to be avoided last thing at night after one year, or it should be followed by rinsing the mouth well with plain water.

Although semi-skimmed and fully-skimmed milks have nutritional advantages for adults they are not suitable as drinks for infants or for preparing weaning foods.

Goats' and sheep's milk are not suitable for infants.

Humanizing cows' milk

To humanize cows' milk it is diluted to make the protein about 1 g/100 ml and lactose or sucrose is added to make the energy value about 70 kcal/100 ml of milk. This is now rarely done at home: it is simpler and safer to use infant formula milk for the first six monthe of life and then to use follow-on formula milk. These milks are fortified with vitamins and minerals. A comparison of breast milk and cows' milk is shown in Table 2.

Other drinks

There is no need to use any main drinks other than infant formula milk, follow-on formula milk and water during the first year of life. Tap water in the United Kingdom is almost always suitable and should be boiled and cooled before use at least up to age six months. If bottled water has to be used it too should be boiled and cooled. Because bottled waters may contain too much dissolved salts to be suitable for infants they are usually best avoided.

Fruit juices, useful sources of vitamin C, usually contain much acid and sugar and need to be diluted before use and fed from a cup and not from a bottle. If teeth have erupted, to avoid damage to them by acid and sugar, juice should be given only at meal times and never at bed time. When given with a meal vitamin C in the juice aids the absorption of iron into the blood. If possible, the mouth should be rinsed with plain water after fruit juice.

The usual soft drinks consumed by older children should not be given to infants. These drinks generally contain too much sugar, artificial sweeteners, acid, salts and sometimes caffeine and other substances unnecessary or undesirable for infants. Tea and coffee are also not suitable for infants.

Diet at 4–5 months

The daily 600 ml of breast milk or infant formula milk should be supplemented by one or more of the following:

a) A sauce made from a mild-flavoured full-fat hard cheese. Soft cheeses must not be given to infants.
b) Some rice-based breakfast cereal containing not more than 3 g of dietary fibre per 100 g of cereal.
c) A smooth custard made with cows' full-fat milk.
d) A smooth purée of well-cooked meat, poultry, fish, vegetables or fruit.

Do not use bread before six months of age as some young infants cannot deal with the protein (gluten) in wheat, rye, oats and barley and they may become sensitized to these cereals, producing a life-long severe intolerance.

There is no need to add salt or sugar to these foods. Feed a little at a time by spoon. Do not add them to the feeding bottle. If the baby rejects the first taste do not try that item again for a few days. Do not try to force the baby to eat an item it does not want. Babies vary greatly in regard to the variety of foods they will accept and the age at which they will accept them. As long as the baby eats a range of different foods, all will be well. Do not overfeed the baby.

Diet at 6–9 months

The daily basic 600 ml of breast milk or infant formula milk should be replaced gradually by follow-on milk. Purées already started should be continued and small amounts of solid food introduced. At around six months infants begin to use their fingers for feeding and this can be encouraged by the use of small pieces of lightly buttered wholemeal toast, hard cheese, well-cooked vegetables and fresh fruit. The yolk of hard-boiled egg may also be used but not the white. Breakfast cereal, preferably rice-based, with a fibre content of about 4 g/100 g of cereal can now replace the low-fibre breakfast cereal. Meat, poultry and fish should be well-cooked and minced or puréed. In addition to the food eaten with the fingers, small amounts should still be fed by spoon. A wide variety of foods can be tried. There is no need to use salt or sugar.

Diet at 9–12 months

During this period there can be a slow change to a routine of three main meals a day with snacks between them, including follow-on milk drinks. Avoid salty or very sweet snacks. Soft drinks are not desirable; milk, diluted unsweetened fruit juice and water are preferable. Biscuits and cake should be used infrequently. Only small amounts of butter or margarine are needed. Meal times are best kept stress-free. Remove all food as soon as it is obvious that the infant is no longer interested in eating it. An infant's appetite will vary from day to day and a healthy infant will eat when it gets hungry and will stop when it has had enough.

The infant should not be allowed to become too fat and food should not be used as a bribe or as a consolation.

If foods of animal origin are scarce high-protein plant foods must be used, such as beans, peas and lentils.

Vegetarian diets

It is difficult to ensure good healthy development of an infant on a vegetarian diet and anybody attempting to do so must seek sound detailed specialist advice from reputable organizations devoted to this way of life.

Fluoride

An adequate intake of fluoride is necessary during the first year of life to enable the developing teeth to become better resistant to decay after they have erupted. In places where the water is low in fluoride, supplements need to be given to infants. Before mothers do this they must get advice from a health visitor or a local doctor. Giving too much fluoride can permanently disfigure teeth. Fluoride in the diet is described in Chapter 43.

Vitamins

For the first six months of life a breast-fed infant of a healthy well-fed mother does not need vitamin supplementation. To ensure an adequate supply of vitamin D the mother and baby need to be out in the daylight for an hour or so each day; strong direct sunlight should be avoided except for perhaps 15–20 minutes.

Infants over six months of age being fed 500 ml or more per day of infant formula milk or follow-on formula milk also do not need extra vitamins because the milks are fortified. If, however, they are getting less than about 500 ml of formula milk per day or are being fed only breast milk or cows' milk then extra vitamins A, D and C are necessary.

For vitamin A, infants aged six months to one year need about 300–1000 μg per day, which is more than the 240–420 μg vitamin A in the average daily intake of 600 ml of breast milk. Hence to cover the needs of almost all infants the 600 ml of breast milk would need to be supplemented by about 700 μg of vitamin A per day. This extra vitamin A can come from either fortified formula milk or from the solid food introduced into the diet during weaning or from vitamin A drops. One μg of vitamin A is equal to 3.3 international units (i.u.).

For vitamin D, plenty of direct daylight on the skin or enough fortified formula milk or vitamin D drops are necessary to prevent rickets. A safe supplementation of vitamin D for infants is 10 μg per day.

The vitamin C in breast milk is normally adequate to prevent scurvy but once weaning has started diluted fruit juice with added vitamin C should be given or vitamin C supplements should be used. A supplement of about 35 mg of vitamin C per day is probably enough.

When giving supplements of vitamins A and D it is important to make sure that there is no overdosage because these vitamins can be very toxic in excess.

Food allergy

When it is known that a member of an infant's family has a definite food allergy, or has a food-related disease, it is advisable to wait until the infant is at least six months old before feeding such foods. Items that may have to be avoided are cow's milk, egg, wheat, rye, barley, fish and particular fruits. If there is need to avoid more than one food, specialist advice should be obtained.

Below three years of age avoid peanuts if from a family with hay fever, asthma, eczema or food allergy.

Children under three years of age should not be given foods containing peanuts (groundnuts) if they come from a family with members suffering from hay fever, asthma, eczema or other allergic conditions.

Chapter 5

Young children (1–6 years)

The general principles of nutrition that apply to adults also apply to young children and are dealt with in the Chapters devoted to the various components of the diet. Those Chapters give the specific needs of young children and also describe what may happen in malnourished children. This Chapter deals in a more general way with other aspects of nutrition in the young.

Snacks

Young children are usually very active and therefore need a plentiful supply of energy, best provided by carbohydrate, though some food rich in fat may be needed to reduce the dietary bulk. Because their stomachs are small, the traditional three to four meals a day have to be supplemented by snacks between meals, the number of these snacks being greater in more active children. In addition, their dietary fibre intake needs to be kept at a modest level so as not to make the diet too bulky. Excess fibre can cause discomfort and even mild diarrhoea.

These between-meal snacks have to be varied to give a good spread of nutrients. Useful snacks are cheese, cheese biscuits, hard-boiled eggs, yogurt, fruit, peanut butter on wholemeal bread, mashed beans and lentils on wholemeal bread, biscuits, cakes and unsweetened orange juice. Up to two years of age drinks of full-fat milk, which may be flavoured, should be used; from 2–5 years semi-skimmed milk may be used. Skimmed milk is not suitable for children under five years old.

A few children are allergic to nuts and must not be given foods containg even very small amounts of nuts as an acute allergic reaction to nuts can be fatal. Children under the age of three years should not be given foods containing peanuts if there are family members who suffer from hay fever, asthma, eczema or food allergy. In addition, whole nuts, especially peanuts, are not suitable for children under five years of age because they may inhale them and choke to death. Care is also needed with pips and seeds.

Salted snacks should be used infrequently and highly sugared foods, although greatly desired, are also not good. This may moderate the desire for salt and sugar in later life. A reduced salt intake may help to keep the blood pressure from rising excessively in those sensitive to sodium and a reduced sugar intake will lessen tooth decay and perhaps also unwanted gain in weight. Fat intake should be moderate but not as low as for an adult. There is evidence that a high fat intake in childhood may predispose to cardiovascular disease in later life. However, fat is necessary for young children because of its high energy content and the need for essential fatty acids. These latter can be obtained from puréed nuts or the use of nut oils for cooking.

Variety of diet

Young children often refuse to try many foods and their diets may seem very boring and limited to adults. Children, however, do not see things that way and provided that the accepted diet has an adequate energy content, with the necessary amounts of the essential nutrients, there is no cause for concern. A healthy normally-growing and developing child is getting a sound diet whatever an adult may think of it.

Breakfast cereals are usually well-liked by young children and with milk they provide a good start to the day; they also make a nutritious snack at any time. The best are well-fortified with vitamins and iron, have a medium fibre content (about 6–9 g fibre/100 g cereal) and do not have a high sugar content. The ingredients should be well-displayed on the ingredients label and the various makes can be compared for their nutritive values. The labels should be read from time to time as the ingredients may be altered without it being obvious.

> Avoid sugary and salted snacks and fizzy drinks.

Supplements

It is desirable that all children from the age of six months to 5–6 years should have their diets supplemented by drops containing vitamins A, C and D. The dose prescribed by the health visitor or doctor must not be exceeded. In addition, allowing daylight (it does not have to be direct sunlight) on the skin will enable the body to produce vitamin D. Any vitamin D not used immediately is stored in the body's depot fat. Never allow the skin to be sunburnt as this may cause skin cancer in adult life

Fizzy drinks

The use of fizzy drinks, widely popular among adults as well as children, can hardly be recommended. They provide very little of nutritive value, the acid and sugar may damage the teeth, they may lessen the appetite and they are moderately expensive. Young children should be taught to ignore them.

Variations in appetite

It may be expected that a child will eat more as it gets older and this does occur during the first year of life. During the second year, however, the rate of growth slows and the appetite therefore may not increase and may even for a while decrease. This stationary or fall in food intake may be a cause of worry and attempts may be made to make the child eat more, which will often result in conflict which the child will usually win. If the mother does succeed, the child is likely to become overweight as the extra food is not needed for the rate of growth at that time. For a child in good health its appetite is the best guide to how much food it needs. The variation in food intake during the second year of life is large: big, very active two-year olds may consume as much as 2000 kcal per day, while small less active ones may take only 1000 kcal per day.

Children's appetites vary greatly.

Vegetarian diets

Many young children are not keen on eating meat, poultry or fish, or only in small amounts. Provided that they eat eggs, milk and cheese, as well as bread, rice, beans, peas, lentils, other vegetables and fruit, their diet will be entirely adequate. The bread or rice should be eaten along with the beans, peas or lentils at the same meal, making the protein mixture of the meal equivalent to that found in meat, poultry and fish. An egg or some cheese or a glass of milk (flavoured if necessary) with such a meal will give added nutritive value. In order to be sure that the iron intake is sufficient, a generous helping of a breakfast cereal which is well-fortified with iron should be eaten at least once a day.

Bringing-up a young child on a diet entirely devoid of animal products, a vegan diet, is difficult and anybody wishing to use such a diet should seek specialist advice and have the child's progress checked at regular intervals.

Additives

Many processed foods contain additives. Some are useful, while others, such as colours, are unnecessary from a nutritional viewpoint. Study the ingredients labels on processed foods and try to restrict the choice to the products with the fewest colours. Almost all additives are probably harmless in the amounts used, but children do on rare occasions react adversely to an additive. Seek medical advice if a problem seems to be present.

Meal times

Keep meal times unemotional.

Meal times should be kept unemotional and not used for quarrels and disputes. Do not try to force a child to eat any particular item of food. If a child regularly rejects a particular item, use something different but of similar nutritive value next time. If that is difficult, try flavouring the rejected item or cooking it differently. Do not let the child use its rejection of food as a means of manipulation and do not use food as a means of manipulating the child. Remember that a healthy child is very unlikely to starve itself sufficiently to do harm: when it gets hungry enough it will eat. Praise a young child frequently and scold it rarely.

Young children tend to like their food to be soft, varied, recognizable, warm rather than hot or cold, not strongly flavoured and served in small amounts. They do not like it to contain inedible items such as gristle or bones and often do not like visible fat. They find eating easier when the plate, cup, glass, spoon and fork are child-size and not adult-size. Very young children often want to explore the food with their fingers before eating it: this has to be tolerated.

Chapter 6

Adolescents (10–20 years)

Adolescence occupies the years between childhood and adulthood, starting at around 10–11 years of age for girls and 12–13 years for boys and ending at around 18–20 years for both sexes. The start of adolescence is related more to total body weight than to age and seems to be initiated when the weight reaches about 30 kg (66 lb), at which point there is a very marked increase in height followed in six months to one year by a marked gain in weight. This is accompanied by the changes needed for development of sexual maturity. The accumulation of body fat seems to be critical for girls and if this is less than about 22 per cent when the weight reaches about 46 kg there is likely to be delay in the onset of menstruation (menarche). The rapid increase in height and weight lasts 3–4 years and is followed by a period of slower growth. The general body changes are usually over by about 20 years of age but bones go on getting heavier for another 3–4 years provided nutrition is good and there is regular exercise, especially lifting and carrying. This period of continued bone growth requires much dietary calcium and is particularly important for women because the more calcium they lay down in their bones during these years the less likely are they to suffer from post-menopausal osteoporosis.

The rapid growth during adolescence requires a considerable increase in food intake which under normal conditions is governed by increase in appetite, resulting in the eating of larger meals and frequent snacks between meals. This between-meal eating, often of sugary foods, is likely to be accompanied by an increase in dental caries unless the teeth are frequently flossed and then brushed with a fluoride toothpaste.

During adolescence there is a change from dependency to independency, which is often shown by a discarding of old ways in favour of new ones, perhaps acquired from more dominant adolescent friends. This may result in good dietary habits being replaced by poorer ones but the reverse may also occur. If dietary advice is given not too often, calmly and with clear reasons as to why the advice is good, the outcome will usually be tolerable. A survey by questionnaire was carried out in 1989 by *Mizz* magazine in collaboration with the National Dairy Council into the dietary knowledge and habits of *Mizz* readers. All the 1000 or so replies were from females, mostly 14–18 years of age, living mainly in England. Even though the respondents were self-selected and may not be truly representative of adolescents as a whole, the answers nevertheless showed that not all adolescents subsist on a nutritionally disastrous diet. The majority of the respondents had a good knowledge of nutritional principles and tried to apply them, which was not always easy.

Energy

This will usually be governed satisfactorily by the appetite, though excessive weight gain should be prevented, not only by eating less but also by taking more exercise. Physical exertion is as important for girls as for boys. The amount of physical exertion undertaken during adolescence varies greatly, so that the need for energy of some adolescents may be as much as twice that of others. But even the most lazy adolescent will notice a need for extra food because of the height and weight spurt taking place.

Protein

Adolescent growth and tissue maintenance is adequately provided for by an intake of 1.5 g protein/kg body weight/day. This is supplied by a diet containing generous portions of meat, poultry, fish, eggs, cheese and milk. It is also possible to get enough protein by mixing pulses (beans, peas and lentils) with cereals (wheat, rice, maize, soya, rye and oats) at each meal. To make sure that dietary protein is used for growth and not as an energy source, sufficient carbohydrate, with or without some fat, must be eaten at the same time as the food containing the protein. Eating mainly the protein at one meal and the carbohydrate and fat at another is not satisfactory because the protein is then likely to be used for energy.

Carbohydrate and fat

Although all the extra energy needed during adolescence can be taken as carbohydrate in cereals, pulses and other plant foods, some extra fat is usually required to keep down the volume of the food and to make it palatable. The amount of extra fat in the main meals should be kept as low as possible because the numerous snacks eaten by adolescents between meals tend to be high in fat.

Calcium

During adolescence the marked growth of the skeleton greatly increases the need for calcium and for many adolescents this need is not completely met. The exact amount of calcium required each day during this period is not known and

the values suggested by various expert committees differ considerably, some advising only 800 mg/day, while others think as much as 1500 mg/day, or even more, may be required. Almost all estimates have been considerably increased since 1988. It would seem sensible to aim at an intake of about 1500 mg of calcium/day. The only normal items of the diet with enough calcium to make an intake of 1500 mg/day feasible are milk and its products; 600 ml (one pint) of milk has about 700 mg of calcium and 100 g of hard cheese has the same amount. In order to keep the daily fat intake from being excessive the milk should be skimmed or semi-skimmed and the cheese should be of a low-fat type. As many adolescents do not wish to drink milk or eat milk puddings a calcium intake of 1500 mg/day is hard to achieve unless calcium tablets are taken. It is best not to exceed 2000 mg of calcium/day because more than that may lead to renal damage in some people, especially if their daily vitamin D is high.

A good intake of calcium and iron is very important for adolescents, especially girls.

Iron

The rapid bodily growth during adolescence requires a plentiful supply of dietary iron, partly because all cells of the body need iron and partly for the formation of extra haemoglobin needed for the increased number of red blood cells in the expanding blood volume. The usual mixed diet in the United Kingdom supplies about 10–15 mg of iron/day for an intake of 1800–2500 kcal, which is often exceeded by adolescents, so that in the absence of food fads most adolescents get enough iron. However, some do not: a 1993 survey of British children aged 12–14 years found that about 4 per cent of the boys and about 10 per cent of the girls were mildly anaemic. Vegetarian women, particularly if menstruating, often do not get enough dietary iron without taking iron supplements and about 25 per cent are mildly anaemic. It should be remembered that the extra milk and cheese recommended for adolescents contain very little iron.

Zinc

Zinc in the ordinary diet is related to the protein intake, 100 g of protein providing an intake of about 15 mg of zinc. Young adolescents need about 10 mg of zinc/day and older ones need about 15 mg. This means that some adolescents do not get sufficient zinc and when younger ones are given zinc supplements they sometimes show an extra spurt in growth. Vegetarians do not show any greater deficiency of zinc than do non-vegetarians because they can aquire the metal from whole-grain cereals, though fruits and other plant foods are generally low in zinc.

Iodine

Iodine is widely spread in food but its abundance varies greatly depending on where the food was grown. As foods in the United Kingdom come nowadays from widely separated places there is less iodine deficiency, though it still occurs. In some people iodine lack causes a simple enlargement of the thyroid gland, which can be seen as a smooth swelling over and around the larynx (a simple goitre). In areas prone to simple goitre the thyroid gland often starts to swell when the need for more iodine occurs at the onset of adolescence. Not all simple goitres, however, are due to iodine lack. Most people with a simple goitre do not show signs of deficiency of action of the thyroid. A daily intake of 3–4 μg iodine/kg of body weight prevents simple goitre.

The best normal source of dietary iodine (Table 3) in the United Kingdom is seafood, with useful amounts coming from meat, poultry, eggs and most vegetables. Eating seafish 3–4 times a week probably suffices for iodine need. Fruits in general are poor in iodine. Iodized salt should be used for cooking and at the table and by food manufacturers who add salt to their products.

Vitamins

The spurt in growth in adolescence necessitates an increase in the intake of food, the metabolism of which requires an increase in the intake of all the vitamins. Most of this additional vitamin requirement is present in the extra food eaten. Not uncommonly, however, adolescent diets are found to be too low in vitamins A, C, B_6 and in folate and supplements of these vitamins may be necessary.

Table 3 Iodine content of food groups

		Iodine in 100 g edible portion
		μg
Seafood	usually very good	65
Vegetables Meat, Poultry Eggs	moderate	30
Dairy produce Cereals	useful	10
Fruit	poor	4

Most adolescents engage in sufficient outdoor activity to acquire adequate amounts of vitamin D by the action of daylight on the skin. In the absence of adequate sunning, however, vitamin D supplementation is needed, either from well-fortified breakfast cereal or given with calcium in tablet form.

Obesity

Being moderately overweight is common amongst adolescents, for boys as well as for girls. Actual obesity also occurs. The urge to eat during adolescence is often very strong and unless the subject learns to stop eating as soon as the appetite is satisfied there is likelihood that over-eating will become an established habit. If, at the same time, regular exercise is neglected the gain in weight can be quite rapid. In addition, adolescence can be a time of unhappiness and eating may be used as a consolation, making weight control difficult.

Chapter 7

Ageing

A child born in the United Kingdom or the USA in 1850 had a life expectancy of about 40 years; a child born in either of these two countries in 1997 can expect to live to be 75–80 years. This longevity may well be exceeded in the forseeable future. As the number of older people grows it becomes ever more important that they should be in good health. This would not only enable them to enjoy life and be useful in their communities, it would also reduce the cost of health care and general maintenance. Everybody therefore has, or will have, an interest in the problems of ageing.

Although almost all the organs of the body show a decline in activity with the passing years, the ability to eat and to digest and absorb food seem to show little deterioration in old people in good general health. Relatively little is known about the nutritional needs of older people compared with the vast amount known about the younger groups. Early signs of malnutrition are generally non-specific and are easily thought to be due to the numerous degenerative changes occurring in old age and it is only when extreme malnutrition occurs that it is likely to be recognised. The early signs may be weakness, loss of weight, anaemia, poor vision, thin skin, easy bruising, fracture of bones, mental disturbance and other changes all of which may be due to diseases occurring more often than does malnutrition. Disease and dietary deficiency may, of course, occur together, especially when the disease in some way limits the ability to maintain an adequate diet.

Although it is known that keeping rats very thin can substantially increase their life span it is not possible to say whether this would be so for humans. Mild overweight does not seem especially hazardous in humans but obesity certainly does reduce the average age at death. It is very difficult to determine the long-term effects of minor differences in diet and body weight in people because in addition to these variations there are usually differences in life-style, diseases and accidents. The most obvious effect of inadequate amounts of food in the early years, especially if poor in protein, is a reduction in life expectancy, particularly if such poor nutrition continues for several years.

Nutritional requirements of elderly people vary much more than those of young people.

The aged are a far more varied group of people than are the young and the middle-aged. Each young person is mostly very like another young person in nutritional needs and other bodily functions. With the passage of time, however, because of genetic make-up, diseases, accidents and life-style, the bodily functions of people become more and more varied.

Protein

The average protein requirement of old people is about the same as for young adults but there is a much wider variation in requirement in the aged. In the absence of disease it is probably best to aim at a high intake of protein so as to be sure to cover the needs of those whose efficiency of protein use is diminished. About 100 g of protein per day, mainly from meat, poultry or fish, should suffice.

When a diet is lacking in protein, body protein is gradually broken-down to compensate for the deficiency. Skeletal muscles become smaller and weaker and the skin becomes thinner. If there is inadequate intake of total energy as well as of protein, there will be loss of weight. If, on the other hand, total energy intake is high rather than low, fat may be laid down, so that the body weight does not fall even though muscle tissue is being lost because of the inadequate protein intake and the person becomes weaker. Because food rich in good quality protein is relatively expensive, many older people live on a diet too low in protein and high in carbohydrate and fat. Their weight commonly increases as their muscles waste and they become less and less active. Merely weighing old people is not a suitable way of deciding whether their diet is a good one.

Fat

As for the young, the fat intake of the aged should be as low as is palatable, with at least half being polyunsaturated. This will probably help prevent any existing atherosclerosis from getting worse, although it may not reduce that which already exists. The fat in the blood after a fatty meal remains high in the older person for much longer than it does in the young. To prevent this fat concentration in the blood from being continuously high in the aged it is better for them to have only one meal a day which contains an appreciable amount of fat. The other meals should be made up of unrefined carbohydrate and protein, with little fat. Such a diet is a healthy one at any age.

Carbohydrate

> Dental health is very important for good nutrition in the elderly.

The metabolism of carbohydrate may become less efficient with advancing years and it is preferable to eat the daily carbohydrate intake in several small meals rather than in three larger ones. Doing this reduces the need for insulin and lowers the blood cholesterol level. Frequent meals, however, especially if rich in carbohydrate, are likely to cause damage to the teeth unless they are kept very clean. Care of the teeth is of great importance in the aged, who may already have poor dentition and any further problems may well be the cause of subsequent poor nutrition.

Calcium

Calcium retention gradually gets worse after about 50 years of age, being more marked in women than in men. This is due to reduced absorption of dietary calcium into the blood, plus increased secretion of calcium into the intestinal lumen and its loss in the faeces. In addition, lying in bed for excessive periods increases loss of bone calcium via the urine. The end result is often osteoporosis with easily fractured bones, and is worse in people who in their earlier years had an inadequate intake of calcium, vitamin D and lack of physical exertion.

It is difficult to increase the amount of calcium in older bones but in some people the further loss of calcium can be prevented or slowed down by raising the dietary calcium to about 1500 mg/day, giving vitamin D supplements either in fortified food or by tablets and by encouraging physical activity (not always easy in old age). Very useful are low-fat cheeses and skimmed milk fortified with vitamin D. One pint (600 ml) of cows' milk contains about 700 mg of calcium, as does 100 g (3.5 oz) of hard cheese. For women in the early post-menopausal years, medically supervised treatment with female sex hormones will greatly improve calcium retention and help prevent bone fracture.

Iron

Little is known about the effect of age on the utilization of dietary iron. Iron deficiency anaemia in the aged should not be presumed to be due to nutritional lack until investigation has shown that there is not another cause. A large helping of an iron-fortified breakfast cereal is a good way to start the day, especially if not much meat, poultry or fish is eaten.

Vitamins

In the absence of disease the aged seem to utilize dietary vitamins well and do not show general evidence of vitamin deficiency, although it has been claimed by some investigators and denied by others that there is poor absorption of dietary vitamins A, B_1 and B_{12}. However, there appears to be no need for extra intake of these three vitamins by older people consuming a good diet.

Some older people have an inadequate intake of vitamin C and folate and for them it would be wise to take one 50 mg vitamin C tablet daily and they should increase their folate intake by eating more dark-green vegetables or a folate-fortified breakfast cereal.

For those older people who do not often go out of doors a daily supplement of 2 μg of vitamin D in tablet form would be beneficial.

Energy

The need for energy usually declines with age because of reduction in exercise, fall in skeletal muscle mass and diminished production of thyroid and adrenal hormones. Some older people, however, do retain most of their muscle mass and continue to be very active, so that their energy requirement does not decline appreciably. Indeed, some people in their sixties and even seventies expend more energy than some people half their age. This considerable variation in energy need by older people means that the amount of food required by the active can be twice that of others, which is not always understood and may create a severe problem where groups of older people are fed communally and given roughly equal amounts. Older people should be allowed to help themselves to the amount of food they need.

Social factors

Many older people, despite a lifetime of work, are poor and are unable to afford a suitable diet, especially of good quality protein, fruit and many vegetables. They tend to live on diets of mainly fat and refined carbohydrate. They may lack suitable cooking facilities, have no freezer and perhaps even no refrigerator, which prevents them from preparing and storing proper meals. Loss of teeth and gum disease, as well as ill-fitting dentures, may aggravate matters for them. Many may need medicines which may reduce their appetite. General infirmity and poor vision often make shopping difficult and reading labels on cans and packages may sometimes be impossible. If they become isolated and depressed any interest in preparing decent meals may vanish. Some succumb to alcoholism.

For the elderly who cannot shop or cook, a hot meal delivered daily to their home may make a profound difference to their nutritional state and hence to their physical and mental health. As well as the delivery of a daily hot meal, an evening cold meal and the next day's breakfast should also be provided. The delivery of the food gives an opportunity to make sure all is well and to report if anything seems amiss. Those who can afford it may pay for this service.

If a few elderly people who would otherwise be alone can meet daily for a communal meal, sharing the shopping, the cost, the meal preparation and the clearing up, there is motivation for a varied diet. Such groups will also give emotional and material support when needed.

In some towns special restaurants for older people provide very pleasant places where nutritionally sound meals are sold on a non-profit basis. These restaurants may also sell cooked food to take away.

Chapter 8

Illness

High energy intake during illness is of paramount importance. Protein intake should be kept high.

During any illness the most important thing after water is the supply of energy. If the subject can eat, as much food as can be tolerated should be given, even if what is eaten would not seem to be a suitable diet under normal circumstances. Even if a person is overweight high energy intake during an acute illness remains paramount. If there is a high temperature even more energy intake is needed and this is further increased if there is restlessness. The need for a high energy intake cannot be over-emphasized and, without it, body protein loss, which probably cannot be completely prevented in an acute illness, will be greater than it need be. If there is inability to eat, or to eat enough, specialist attention is essential.

Protein

All moderate to severe illnesses cause a loss of body protein, mainly from skeletal muscle, and a fall in body weight. If an injured person takes no food during the first three days and expends 6000–8000 kcal in that time, to which protein contributes 20–25 per cent (1200–2000 kcal), the lost muscle, which is about 80 per cent water and 20 per cent protein, will amount to about 1200–2000 g (2.7–4.5 lb). This is unlike the situation in simple starvation, when mainly body fat is lost. The amount of protein lost depends on the severity of the illness and its nature, including the amount of pain suffered. Fractures of long bones and extensive skin burns are particularly likely to cause much protein loss. Prolonged bed rest also produces loss of protein because of atrophy of skeletal muscle. The protein is lost partly because the disease process may directly destroy body tissue and partly, usually mainly, by the action of hormones secreted by the adrenal glands. These hormones are produced in large amounts when much tissue has been damaged and pain inflicted and their action rapidly breaks down body protein; they are known as catabolic hormones. Those that build up body tissue are the anabolic hormones and are secreted subsequently during the recovery phase. Break-down and production of tissue can occur simultaneously. When loss of body protein is large it may be accompanied by weakness, poor healing, lowered resistance to infections and may lead to unexpected death, especially if there has been malnourishment just prior to the illness.

During the recovery period there is a marked need for protein of high biological value (see Chapter 18). In an adult in good health relatively little protein is needed each day for general tissue maintenance, but during recovery from illness the amount of protein needed becomes similar to that of a rapidly growing young child. There must also be extra water intake for the renal excretion of the waste substances produced by the increased protein metabolism. A urine output of at least 750 ml/day is desirable if the kidneys are healthy.

There are two groups of diseases during which a high-protein diet is undesirable: these are diseases in which liver function is badly disturbed and diseases in which there is severe kidney damage. Under these conditions it is necessary not to feed more protein than the liver can metabolise and to make sure that the end products of that metabolism can be excreted by the kidneys or removed by dialysis. In subjects who have lost considerable body protein it may be necessary to attempt to replete them before undertaking surgery or debillitating therapy (cancer chemotherapy; radiotherapy).

Most patients are emotionally stressed, which may cause loss of appetite (though rarely some may eat too much). Thus it is essential that extra food given to them should be acceptable as well as being nutritionally sound.

Vitamins

During an acute illness there is need for increased vitamin intake, especially for the vitamin B group and vitamin C. The exact amounts required are not known but it is believed that as much as five times the normal intake is desirable for at least vitamins B and C. A large intake for a limited time, even if not needed, would be harmless.

Medicinal drugs

Medicinal drugs can cause a nutrient deficiency by reducing appetite, inducing nausea or vomiting, by interfering with the intestinal uptake of specific nutrients or by preventing the proper action of essential nutrients at the tissues. Such actions can lead to serious deficiencies of vitamins and minerals. Patients on drugs known to have such effects must be closely monitored so that deficiencies can be corrected, especially if the drugs are taken for long periods.

Exclusion of nutrients

In some conditions it may be necessary to exclude certain nutrients from the diet or to provide them in only limited amounts. For example, in the condition known as phenylketonuria (PKU) it is essential to carefully limit the dietary intake of the essential amino acid phenylalanine as soon after birth as possible; in lactose intolerance the sugar lactose (present in milk) needs to be kept at a low level; in diabetes mellitus glucose and sucrose intake must be carefully controlled; in conditions in which dietary fat is poorly absorbed by the intestine the fat in the diet must be kept low; in patients with excessive body iron the dietary intake of the metal has to be restricted; when the blood calcium levels are abnormally raised calcium in the diet needs to be lowered; in obesity the energy content of the diet should be diminished; when there is excessive water in the body and the kidneys are unable to excrete it, water intake has to be limited; in liver and kidney failure dietary protein and perhaps sodium have to be very low; in high blood pressure reducing the salt eaten may be of value.

Chapter 9

Anorexia nervosa and bulimia

This condition usually afflicts bright, able, intelligent, often well-educated obsessional women of around 16–25 years of age who see themselves as too fat. Men may suffer this disorder but the women far outnumber them. At the beginning they may indeed be slightly plumper than average but they are rarely overweight. In an effort to become what they believe is a desirable weight they begin to eat less and this at first causes little alarm among their families and friends because being thin is generally fashionable for young women. In contrast to the people who are actually overweight, for whom losing weight is really necessary and who usually talk and complain a good deal about their troubles, anorexics mostly say nothing about their dieting and may even go to some effort to conceal the fact that they are eating less. They therefore often eat mainly or only in private. Later, when they have lost enough weight to worry their families they maintain that they are still too fat and may even pinch up a piece of wasted flesh to demonstrate how fat they are, although they are obviously alarmingly thin. It is a remarkable disorder of perception and rational discussion is almost always fruitless. By this time, when they weigh perhaps only 35–40 kg (80–90 lb), there is likely to be some anaemia and in women menstruation will have ceased. Such people are usually irritable and emotionally isolated, having given up most or all of their social life and perhaps even their job. They may now see themselves as of a desirable size or may even yet believe that they are still too fat, continuing to lose weight until they are in grave danger of dying. Family tensions at this stage are usually running high and psychotherapy will probably be needed for everyone with whom the anorexics are living.

It is amazing how far advanced the anorexic condition becomes before medical help is sought. Treatment consists of specialist re-feeding, often in hospital, and psychotherapy. Only about half recover fully. Of the others, about 5 per cent die of starvation and the rest live out their lives only partially back to normal. The cause of anorexia nervosa is unknown.

Bulimia

In this variation of abnormal perception of body image, the subject, once again usually a young woman, tries to control body weight by inducing vomiting. This is sometimes preceded by binge eating. The periods of binge eating and vomiting are often followed by periods of reduced food intake which in some cases may be indistinguishable from anorexia nervosa. Body weight loss in bulimia is less likely to be as drastic as that in anorexia nervosa and therefore less life-threatening. The cause of bulimia is unknown.

Chapter 10

Vegetarianism and veganism

People who live on a diet totally devoid of animal products are known as vegans and are rare in the western world. Much more common are those who eat milk, milk products and eggs in addition to plant foods: they are known as lacto-ovo-vegetarians, or vegetarians for short.

Vegetarians are not, of course, a homogeneous group and among them are people who further restrict their diets in various ways. Some of these extra restrictions may be harmless, although they may make it very difficult to devise a nutritionally adequate diet; some of the extra restrictions, on the other hand, can be actually harmful. When considering the health of vegetarians in comparison with the health of the omnivorous general population it needs to be remembered that vegetarians often vary markedly from the general population in their use of alcohol, tobacco and other drugs and in their attitude to a healthy way of life.

Protein

As will be described in Chapter 18, mixtures of plant foods are valuable sources of protein. When eaten together, pulses (beans, peas, lentils) and cereals (wheat, rice, soybeans, rye, barley) provide a protein source as good as animal protein. In addition, eggs, milk and milk products have proteins of high nutritional quality and the amount of protein in hard cheese may equal or exceed that in animal muscle. Nuts and seeds, when puréed, are also useful sources of protein. Thus it is easy for a vegetarian to have an entirely satisfactory protein intake and it is not difficult for a vegan, though the latter has to eat a bulkier meal. The daily protein requirement remains the same for a vegetarian on a good diet as for a person eating all types of animal foods. The protein content of common foods is given in Table 5 (see p. 81).

Fat

The fat intake of vegetarians is somewhat lower than that of the general population. Currently, for vegetarians it is about 38 per cent of total energy intake, whereas it is about 41 per cent for the average omnivore. The vegetarian diet is higher in polyunsaturated fatty acids than the general diet and is lower in saturated fat so it should thereby convey a health advantage.

Dietary fibre

Vegans have a daily intake of about 50 g of dietary fibre, for vegetarians it is about 40 g and for omnivores it is about 25 g. Although a diet high in fibre is generally beneficial, an intake as high as 50 g per day may at first give a sense of bloating and will certainly produce considerable amounts of colonic gas which may cause discomfort. In addition, the fibre may reduce the intestinal absorption of iron, calcium and zinc and perhaps other minerals. The large quantity of plant material eaten may increase the phytate intake which will also lessen the absorption of iron and calcium. Severe mineral deficiency in vegetarians, however, is rarely seen.

Calcium, zinc and iron

For vegetarians, the dietary supply of calcium is very high because of their intake of milk and cheese, whereas for vegans the calcium intake may be less than desirable, especially in pregnancy, in lactation and in children and adolescents. Furthermore, vegans are likely to absorb calcium poorly because of the very high fibre and phytate in their diet. Many vegans would probably benefit from dietary supplementation with calcium tablets.

> A vegetarian diet usually requires more iron, calcium and zinc than does a diet containing meat.

The zinc intake in vegetarians, as in omnivores, is very variable and their zinc status is uncertain. The large fibre, phytate, oxalate and tannin intakes by vegetarians will tend to limit their absorption of both zinc and iron.

The intake of iron by vegetarians and vegans is usually high, but the iron in plants is less easily absorbed than is the iron in animal foods. Nevertheless, most vegetarians and vegans are not iron-deficient. Their high vitamin C intake will aid the uptake of iron by the intestine.

Vitamins

Apart from vitamins B_{12} and D, vegans and vegetarians can get all the vitamins they need from plant foods. Vitamin A, which is found only in animal foods, can be derived from β-carotene and similar pigments widely spread in plants;

vitamins of the B-group except B_{12} are easily obtained from numerous plants; vitamin C is abundant in citrus and other fruits; and vitamins E and K are available in many plants, with vitamin K being synthesized by micro-organisms in the large intestine from where it can be absorbed into the blood. The need for vitamin D can be satisfied by sunning the skin of the face and hands for perhaps an hour or two each day. Further amounts of vitamin D are available for vegetarians from dairy products and for vegans from vitamin D-enriched foods.

Vegans need vitamin B_{12}-enriched food.

Vitamin B_{12} is freely available to vegetarians in eggs, milk and milk products, but for vegans vitamin B_{12} has to be obtained from enriched food or by supplementation. Normally available plant foods do not contain vitamin B_{12}. When the intake of this vitamin is high it can be stored in the liver sufficiently well for a 5-year supply to be accumulated.

Vegetarian children

Vegetarian and vegan children can grow and develop normally. It is essential that the mother is well-fed during her pregnancy and lactation and that her mineral and vitamin intakes are high. During infancy most vegetarian children do well because most are breast-fed for at least six months. If bottle-fed, vegan infants can be fed on an infant formula based on soya, enriched with iron, calcium and vitamins. After six months of age, all vegetarian and vegan infants must have supplements of iron and vitamins A, B_2, B_{12}, C and D. These supplements may be started at one month, with vegan children getting extra calcium.

Once weaning starts it is essential that vegetarian and vegan infants and young children get an adequate energy intake by keeping the fat in the diet higher than would be necessary for omnivore children. This is because the vegetarian diet is bulkier than the omnivore diet and there is the danger that the small stomachs of infants and young children will not be able to accept a sufficient volume of vegetarian food, leading to an energy deficiency. That this is a real problem rather than just a theoretical one is shown by the finding that many vegetarian children are lighter than age-matched omnivore children, although this does not necessarily mean that the vegetarian children are unhealthy. Most vegetarian children do not show signs of dietary deficiency because their parents know how to prepare good vegetarian diets.

General health

Vegetarians are lighter in body weight and have a lower body mass index than omnivores and because of the low incidence of obesity among vegetarians there is a low incidence of late-onset diabetes mellitus. Most life-long vegetarians and vegans have a lower blood pressure than the general population. In addition,

Vegetarians are just as healthy as omnivores, though usually lighter.

coronary heart disease is less common in vegetarians than in omnivores, which may be due to lower body weight, to increased intake of dietary fibre, to lower blood cholesterol levels, to lower saturated fat intake, to higher intake of polyunsaturated fatty acids, to an increased intake of vitamins A, C and E and to a generally healthier life-style. The high fibre diet also seems to reduce the risk of developing some large bowel diseases. Despite all these advantages apparently derived from their diet and general regard for their health, vegetarians do not appear to live longer than non-vegetarians, although they may be healthier while they are alive.

Chapter 11

Diet selection

What to eat on a weekly basis

Diet selection based on a rigid daily routine, not always possible for many people, is not necessary. Provided that there is an adequate intake of the essential nutrients every few days all will be well, even for substances which the body hardly stores, such as vitamin C and protein. For some items, such as iron and vitamin B_{12}, which are very well stored, quite lengthy periods of low intake may do no harm. Sensible eating should include keeping to a good basic routine and avoiding alternating periods of marked over-consumption with periods of severe under-consumption.

Many dietary plans have been devised over the years. They vary a lot, but the message of the sound ones is that the best diet is a varied one, with many different types of food being chosen. A healthy habit is to eat an adequate amount of protein each day, supplemented by enough fibre-rich cereals, pulses and vegetables to provide the bulk of the energy needed. Only a modest amount of fat is required. To this should be added two or more items of fruit each day. Such a diet will give the energy required plus all the essential substances needed for health even if chosen by a non-expert. The variety of the foods that can be used is very large and which items are eaten will depend on personal preference, availability, price and social and religious customs. If alcohol is used it should be taken in moderation with at least one alcohol-free day each week.

For vegans the task of designing a suitable diet may be difficult and they usually need specialist advice.

The various Chapters of this book give accounts of why nutrients are needed, how much is needed by different groups of people, in which foods the nutrients are most abundant and what may happen if things go wrong.

When buying food be sure that it is in good condition. Frozen food kept sufficiently cold may be better than unfrozen food that has been on a shelf at room temperature. Read the information boxes on food packages and compare different brands of the same item before making a choice. If there is no nutritional information given, try to avoid that product. Preservatives are not always undesirable and may indeed be necessary to keep some foods from spoiling quickly. Added colours are almost always unnecessary.

When selecting items for a meal, eye-appeal is important. Have several different colours on the plate and have a variety of textures, some crunchy, some softer and moist. Contrasting flavours always make food more interesting. Use salt sparingly, preferably iodized. Hot food is not nutritionally better than cold food, although hot food may have a more interesting taste. A bright uncluttered table with attractive tableware will add to the enjoyment of food, as will an unhurried and calm atmosphere.

Highly nutritious and attractive meals can be prepared at modest cost and quickly: there is little to be gained nutritionally in expending much money and tedious effort.

Recommended Dietary Amount (RDA)

The RDA indicates how much of a nutrient a person needs. It is given as a daily value but it is quite safe to take more or less on any particular day provided that the intake over several days is at least the RDA. It is not an absolute value but more of a guide and it has not been determined for all known nutrients. The values given are generous ones so as to meet the requirements of almost everybody. During periods of illness it may be desirable to exceed the RDA. For persons purposely losing weight it is essential to be sure to reach the RDA for protein, vitamins and minerals, which may require the taking of supplements.

There has been a series of new definitions for nutrient requirements and the term Reference Nutrient Intake (RNI) is replacing the RDA. The RNI is the Estimated Average Requirement (EAR) plus two standard deviations, which means that the RNI will satisfy the needs of 97 per cent of the population. This book, however, uses the RDA values because they seem sufficiently satisfactory, will be found in most reference tables and are still used in nutritional information boxes on food packages.

The tables in this book give general rather than very detailed information, which should be quite adequate for day-to-day use.

Chapter 12

How to interpret food labels

Ingredients

> The law requires all ingredients to be listed, with the most abundant substance listed first.

All the ingredients of a processed or manufactured food must be given in a box marked 'Ingredients'. The method of giving this information is governed by law: all the ingredients must be given together in full except for the group described as flavourings, which need not be named separately. The law further requires that all the ingredients be listed in order of weight, the most abundant substance being named first, followed by the next most abundant item and so on, with the least abundant substance given last. All the additives used in processing this particular food must also be given, naming the category of each additive and giving its number or chemical name or both. If an additive is permitted by the European Union its number is preceded by 'E'. Additives used in processing the components of a food need not be given and this might at times be of major importance to a consumer. For example, if flour is bleached during its manufacture, a second firm making perhaps biscuits from this flour would not have to itemize the bleaching agent even though some might be present in the biscuits. Such indirectly present additives may be omitted from the ingredients box.

Any information given outside the ingredients box need not be complete. Thus outside the box there might be the statement 'Stewed meat' plus a further statement saying 'Contains beef'. This does not mean that the package contains only beef: it means that there is some beef in the product but that there may also be one or more other types of meat as well as non-meat material, water, preservatives, flavourings, colours and other substances. The part of the label outside the ingredients box merely indicates the type of product being offered and is not usually a complete description of the product. Further, the word 'beef' means that any part of a cow may have been used, not just steak unless the word 'steak' is itself used. Similarly the word 'pork' means that any part of a pig may have been used, not just ham unless it says 'ham', and the word 'chicken' means that any part of a chicken may have been used unless it says 'chicken breast' or 'chicken leg meat'. Other statements may also be on the label outside the boxes but they must not make overt health claims unless they satisfy the rules governing medicines.

Water naturally present in a food need not be listed but if extra water is added it must be itemized if the amount of added water is more than 5 per cent of the product.

A typical ingredients box for wholemeal bread might look like this:

Ingredients

Wholemeal flour; water; yeast; salt; wheat gluten; vegetable fat; emulsifiers: E472(e), E471; soya flour; flour treatment agent: E300 (L-ascorbic acid).

Wheat gluten is a protein derived from wheat. The emulsifiers E472(e) and E471 are mixtures of glycerides (fat components). The flour treatment agent L-ascorbic acid is vitamin C.

Nutritional information

Many food manufacturers are now giving detailed nutritional information about their products. This information is usually given in a box similar to the ingredients box. The analysis almost always shows the values for 100 g of the food and in addition sometimes gives the values for average-sized servings. A typical example for wholemeal bread might look like this:

Nutritional information	Typical values per 100 g	Per slice
Energy	900 kj (214 kcal)	330 kj (79 kcal)
Protein	10 g	3.50 g
Carbohydrate	40 g	13.6 g
of which sugars	3 g	1 g
Fat	3 g	1 g
of which saturates	0.8 g	0.3 g
Fibre	6.5 g	2.2 g
Sodium	0.40 g	0.15 g

By comparing the nutritional information boxes for different breads it is possible to choose, for example, a bread with a high fibre content (6–8 g fibre per 100 g) or a low one (2 g fibre per 100 g); or a bread giving a high energy yield per slice (80 kcal) or a low one (30 kcal). The detail given for similar foods varies considerably between manufacturers.

These nutritional values are average typical values and it is not useful to discriminate between similar numbers: thus 6.5 g should be considered as much the same as 6.2 g and 6.8 g. Pieces of the same food may often vary a lot in their nutritional content, one slice of meat having more fat and hence less protein than another slice. Nevertheless, good nutritional labelling is essential if shoppers buying food are to be able to choose what they want to eat. Apart from the occasional special item of food, it is sensible to ignore food for which the manufacturer does not care to give nutritional information.

The nutritional value per 100 g enables a comparison to be made between different foods, while the nutritional value per serving tells how much of each component is eaten per meal.

Dated foods

If a food is labelled 'Use by . . .', it should not be used after that date unless it has been kept frozen. If labelled 'Best before . . .', it may be used after the indicated date although it may have deteriorated in taste or consistency.

The current methods of labelling foods are only partially useful. What are needed are clear and more comprehensive lists of ingredients, additives and nutritional values and all foods should be labelled in a similar way so that easy comparisons may be made. When considering the value of a food, decide by reading the ingredients and nutritional information boxes rather than the other parts of the label.

Chapter 13

Food additives

As well as the use of heat and cold in food processing and preservation, there are several thousand substances that may legally be added to food to make it more palatable, more nutritious, safer, more attractive to the eye, last longer on the shelf or to make it cheaper. Apart from a few materials that are permitted because they have been used for very many years (salt, sodium nitrite, sodium nitrate, smoke), most are of relatively recent use and they have been tested for safety in various ways, so that they are unlikely to cause harm to most people. Some additives, however, may be harmful for a very few people, either very soon after consumption or sometimes after being consumed for some years.

In addition to the presence in food of purposely used additives, there may also be substances entering food from packaging. These indirect additives are likely to be present in only very small amounts but they must be taken into account when assessing overall safety.

Other indirect additives may be present in food because they have been used to treat animals or plants from which the food was derived. For example, antibiotics given to animals and pesticides used on plants may be in food, although they have not been purposely added to the food.

In the United Kingdom there are several legal constraints on the addition of non-food items to any food: some of these controls are national and some are of European Union origin. The purpose of these constraints is to prevent anything harmful from being added to food, although that may occur on very rare occasions. Before a new additive is permitted for use it is necessary for the applicants to show that it is safe, that it is an improvement on any existing permitted similar additive, or that it is as good as an existing additive but cheaper (which could benefit consumers).

Not all additives are artificial; many are natural substances often found in food (pectin, vitamin C, lecithin, calcium chloride). Many artificial additives are very similar to natural substances but with slight alterations to their composition to make them more suitable for their purpose. On the other hand, some additives have been specially made to perform a particular task. Some additives are approved for use on only a national basis, while others are given European Union-wide approval. Most additives have been given a number. If the additive has European Union approval the number is preceded by 'E'. All additives must be itemized in the food ingredients box and this may be by number or by its common or chemical name or by its name plus its number. For example, vitamin C can be listed as E300, vitamin C, L-ascorbic acid or as a combination of these. If an additive has no number it must be listed by its name.

As well as itemizing additives by number or name, it is also necessary to indicate the group to which the additive belongs. The main groups are preservatives, antioxidants, emulsifiers, stabilizers, flavourings, flavour-enhancers, sweeteners and colours. Thus when β-carotene is added to a food to enhance the colour it will be listed in the ingredients box as Colour: 160(a), Colour: (160a) β-carotene or Colour: β-carotene.

The use of additives in food is a very complicated as well as a very controversial matter. It is sometimes decried as being entirely undesirable and a demand is made for all food to be supplied only in its natural state. This is unrealistic. For food to be available only in its natural state, with no processing or additives, would severely reduce the choice of safe food for most people, especially away from rural areas. Such food would be available only in season, would be drastically limited in range and would often be gravely inadequate. Food processing is essential if there is to be a safe, varied and balanced diet throughout the year. Furthermore, in many communities few households have enough labour for the preparation of meals from entirely unprocessed food. The immense advantages of processing food far outweighs the small risks inherent in the procedures if these are guided by adequate regulations. In all probability the number of additives could be reduced and perhaps the use of colours could be discouraged except for a few which are in fact valuable, such as β-carotene.

Food additives can be advantageous.

When buying food, it may be better to choose items which contain the fewest additives and no colours but some obvious exceptions to this advice are fortified breakfast cereals, which may have very valuable amounts of vitamins and minerals added to them; bread with an anti-staling agent is more convenient than bread without it, which may last only a day or two; milk with added vitamins A and D is more nutritious than natural milk; foods with added dietary fibre are very valuable when the diet is otherwise low in fibre.

Categories of additives

Antioxidants

The main function of antioxidants is to stop fat and oil in the food from going rancid. They also prevent the deterioration of fat-soluble vitamins. Antioxidants can delay or prevent the discoloration of food which is caused by enzyme browning, thereby making food more attractive and extending its storage time.

Preservatives

These substances enable food to be kept longer and safer, benefiting the manufacturers, shops and consumers. By preventing spoilage they can keep food cheaper and may make the storage of food at home more convenient. In less developed communities preservatives are almost essential.

Colours

> Food preservatives are sometimes essential. Colours are nutritionally unnecessary.

The use of colour in food is probably the most contentious aspect of food additives. It may promote the use of a nutritionally valuable food which would otherwise be ignored. Colour is also used to give a product a consistent appearance when that might differ from time to time. Generally, the use of colour should be kept to a minimum.

There is no colour added to baby food in the United Kingdom.

Emulsifiers and stabilizers

These enable the mixing of oil and water and prevent the oil and water separating during storage.

Sweeteners

Some of these are natural sugars such as sucrose (table sugar), lactose and mannose, while others are artificial sweeteners such as saccharin, acesulfame and aspartame. The natural sugars are usually used when bulk needs to be given to the food, while the much more powerful artificial sweeteners tend to be used when added bulk is not necessary. The natural sugars provide useful energy, while the artificial sweeteners do not. The natural sugars are not considered to be additives.

Flavourings

There are several thousand items in this group. This is necessary because the flavours of natural foods are made up of very small amounts of a very large number of substances. They are used in food in very small amounts and ingredients boxes need indicate only that flavouring has been used without itemizing each individual substance.

Humectants

These prevent foods from drying out. Glycerol (E422) is a popular humectant; it occurs naturally in the body.

Firming and crisping agents

These are commonly used in canned items which might otherwise become mushy. Calcium chloride (number 509) is often used.

Flour improvers

These make dough stronger and are said to improve the quality of bread, cakes and biscuits.

Other common additives stop liquids from foaming; increase foaming; make food look shiny; prevent food being too sticky; prevent food from caking.

Chapter 14

Food allergy and food intolerance

Food allergy

Food allergy is almost always the result of sensitization of body cells by a food protein which has been absorbed into the blood intact. Food protein is normally digested to its component amino acids which are then transferred to the blood, but very minute quantities of food protein can sometimes escape this digestion and traverse the intestinal wall. This undesirable event can occur at any age but is most likely during infancy, which is why foods known to produce allergy (shellfish, egg white, cows' milk, wheat, nuts, fish) should not be fed to infants, especially those with a family history of allergy. If it is necessary to feed these allergy-producing foods to infants the foods should be thoroughly cooked. This changes the proteins sufficiently to make them less likely to cause allergy. Many tissues of the body can become sensitized, particularly the lungs, the skin and the eyes and the allergic reactions may be acute or chronic, immediate or delayed for several hours. Once tissues have become sensitized to a food the change may be life-long or it may slowly disappear over the years if that food is avoided. Tissues can become sensitized to more than one food so that several may have to be omitted from the diet. Among the common illnesses caused by allergy to food are skin rashes, blisters, asthma, conjunctivitis, abdominal pain, diarrhoea and headache. All these conditions, of course, have other more common causes. True food allergies are uncommon and difficult to diagnose.

> Food allergies are uncommon; food intolerance occurs quite often.

Breast milk may contain intact proteins from the mother's diet and a susceptible infant can become sensitized to these proteins. In a family with allergy it is advisable for the mother during lactation to heat thoroughly any foods to which her family are sensitive.

Childhood allergy

Why some children develop a food allergy and some do not is not known, although family histories of allergic conditions strongly suggest that there is an inherited trait. Where both parents have allergic reactions there is a more than 50 per cent chance that their children will also develop one.

The commonest foods to cause allergy in young children are eggs and cows' milk, but this may be because these two foods are fed often and may be insufficiently heated. When thoroughly heated, egg white and cows' milk may be fed without harm. Allergic children often develop eczema and in many of them cows' milk may cause bleeding into the intestine. This intestinal blood loss may occasionally produce anaemia. In some extreme cases, especially in allergy to nuts, a bronchial allergic response may be so great as to cause death by suffocation. Children under the age of three years from a family with hay fever, asthma or eczema should not be given foods containing peanuts.

Treatment of allergy

If the allergic response is immediate there is usually little difficulty in deciding which food caused the reaction, which may come on within a few minutes of eating it. In some cases, however, the allergic response may be delayed for several hours and an accurate diagnosis is then more difficult.

The only satisfactory treatment is to avoid totally the offending item of food on a permanent basis. This can be very hard to do. Persons allergic to nuts need to be particularly cautious because ground-up nuts are often used as fillers in many processed foods (there should be a warning on the label). Desensitization can be tried but it is very rarely satisfactory and can be hazardous.

Antihistamine drugs are sometimes of value, especially for skin events, while drugs which dilate the bronchi are valuable in asthmatic attacks. Other treatments under medical supervision are available.

Food intolerance

This is a condition in which a particular food causes one or more undesirable effects which are not mediated by the immune system. These reactions are not true allergies and they are far more common than true allergies. For example, some foods may cause hyperacidity with acid regurgitation, an exceptional sense of fullness, generalized abdominal discomfort, nausea and diarrhoea. Intolerance to lactose (milk sugar) occurs frequently in non-white adults and is brought about by a lack of a sufficient amount of the enzyme lactase in the small intestine, so that intact lactose enters the large intestine where micro-organisms produce excess gas and lactic acid from the sugar, causing abdominal cramps and diarrhoea. Another example is intolerance to fat in people who cannot adequately digest it, resulting in excess colonic gas being formed and substances which induce diarrhoea. A few people react excessively to caffeine in tea and coffee and may get palpitations and a sense of anxiety.

Chapter 15

Food toxicity

Anyone consuming food in a well-organised developed country in which they have lived long enough to have acquired immunity to the main local micro-organisms is usually confident that the food is safe. If they have any qualms, it is most likely to be about food additives, even though these have been thoroughly tested for toxicity. Apart from the rare people who may be allergic to a particular food additive, these substances present the least threat to health. Everyday food may, however, cause trouble from time to time, most often mild though occasionally serious or even fatal, the toxicity arising from natural poisons in the food or from contamination. The contamination may be with microbes because of poor standards of hygiene, or, far less commonly, some poisonous material may have accidentally entered the food. The purposeful adulteration of food with a harmful material in well-organised countries is now extremely rare.

Many ordinary foods contain some naturally-occurring toxic substances but they are very unlikely to cause harm. This may be because the toxic materials are present in very small amounts, or are not sufficiently well absorbed into the blood, or because they are quickly neutralized by the liver and then excreted in the bile or in the urine. The body's protective mechanisms can, of course, be overwhelmed if the intake of toxic material is large, or if the liver and kidneys are not working adequately. Of the everyday foods eaten in the United Kingdom, beans, potatoes and rhubarb are the most likely sources of trouble from naturally-occurring toxins, but even with these poisoning is very uncommon if simple precautions are taken.

Beans

Most types of beans contain substances called lectins (also known as haemagglutinins). In most cases of poisoning with lectins there is nausea, vomiting and diarrhoea, with abdominal discomfort, shortly after eating the beans (most usually red kidney beans). Raw beans (except blackeye beans and lentils) need to be soaked overnight in clean cold water. After soaking, the water should be discarded and all beans (whether soaked or not) have to be boiled in copious clean water for at least 15–20 minutes. This water must also be discarded and not used for anything else. The beans are then rinsed with clean water, after which they are ready for use. Canned beans do not need this soaking and boiling as they have been prepared by the manufacturer but it is good practice to rinse them before use.

Potatoes

Almost all potatoes contain a toxic substance called solanine, although the amount in well-stored potatoes is insufficient to cause harm. To some extent this is because the solanine is chiefly in the skin and the eyes, which are usually discarded, and a lot of the solanine is extracted into the water when potatoes are boiled. The toxin increases markedly if potatoes are kept in the light, when they begin to sprout and go green. Sprouting or greeny potatoes should be discarded. Mild poisoning with solanine produces abdominal pain, vomiting and diarrhoea; severe poisoning may induce fever, circulatory collapse, hallucinations, stupor and may even be fatal. Poisoning by well-kept potatoes is, however, extremely rare.

Rhubarb and spinach

> Oxalic acid in rhubarb and spinach prevents absorption of iron and calcium into the blood.

Rhubarb stalks and spinach contain appreciable quantities of oxalic acid, which interferes with the absorption into the blood of calcium and iron eaten at the same meal. When the diet has ample calcium and iron this may not matter, but on a diet low in these two minerals the use of rhubarb and spinach is not a good idea. Far more toxic than the stems are the rhubarb leaves, which are heavily laden with oxalic acid and must never be eaten. Mild oxalic acid poisoning causes abdominal pain, vomiting and diarrhoea; severe poisoning from rhubarb leaves can produce convulsions, coma and death. In general, these two foods should be used in only small amounts for children.

Cabbage, Brussels sprouts and broccoli

These foods contain substances which can interfere with the proper function of the thyroid gland, preventing the uptake of iodine and the production of the hormone thyroxine. These toxins are known as goitrogens. Under ordinary circumstances cooked cabbage, sprouts and broccoli are very valuable foods but on rare occasions when eaten raw in large amounts they may produce goitre-like effects, especially if the diet is low in iodine.

Cheese

Some cheeses, particularly the cheddar type, may contain a compound called tyramine, which is normally metabolized to a harmless substance by an enzyme called mono-amine oxidase. People suffering from depression may be treated with a drug (a mono-amine oxidase inhibitor; MAOI) which prevents the action of this natural enzyme, resulting in an accumulation of tyramine in the body from the cheese. A rise in blood tyramine can result in a marked rise in blood pressure, palpitations, severe headache and may even be life-threatening. People taking mono-amine oxidase-inhibiting drugs must be very careful to avoid cheese, even for a few weeks after stopping taking the drug.

Fish

Some species of sea-fish are poisonous, especially those feeding near the surface of the water. Deep-feeding fish are more likely to be safe. Some are poisonous all the year, while others only at special times. For some, the fish muscle may be poisonous, while for others it is only the liver. The poisons are usually derived from the creatures on which the fish feed. This problem also arises from time to time with mussels and similar shellfish which live on plankton. Occasionally during the summer some species of plankton reproduce very rapidly, enough to turn the coastal waters red (red tides). The toxin produced by the plankton is concentrated in the bodies of the mussels and makes them unfit for consumption. The toxin is not usually completely destroyed by cooking. It rapidly causes vomiting, tingling in the limbs and muscular weakness sometimes sufficient to cause death from damage to the muscles of respiration.

Mushrooms

It is extremely unwise to eat any wild mushrooms unless there is certainty that they are safe. Some mushrooms and many other fungi contain toxins that may prove fatal. There are frequently several toxins inducing at first abdominal pain, vomiting and diarrhoea, followed by damage to the liver and the kidneys. It is the effect on the kidneys that is usually life-threatening because of renal failure. Other manifestations of toxicity may also be present, of which hallucinations are well-known: indeed some poisonous fungi have been deliberately eaten because of the psychic effects that they produce.

Peanuts (groundnuts)

If peanuts are allowed to become mouldy they may become infested with a fungus (Aspergillus flavus) which produces a toxin called aflatoxin. This is a very dangerous material because it not only causes general liver damage but it may also cause liver cancer. Peanuts imported into the United Kingdom are screened for aflatoxin but if any purchased peanuts appear to be in less than perfect condition they should be destroyed immediately. Other crops may also become contaminated by A. flavus if they are allowed to be moist and warm.

Liver

For most people liver is a valuable and cheap food. It should, however, be avoided by women who may be pregnant or may soon become pregnant. This is because of the possibility of damage to the developing baby by excessive amounts of vitamin A which may be in the liver. This large vitamin A load is brought about by giving animals fish products very rich in vitamin A. Fish liver itself should never be eaten.

Ergotism

Mouldy rye and also other cereals can be infested with a fungus called *Claviceps purpura*, which produces a powerful poison which causes ergotism. Anyone eating food made from these mouldy cereals may experience hallucinations, disordered movements, a burning sensation in the limbs and a deficiency of blood supply to the hands and feet, which may in severe cases result in gangrene. Although now rare in the United Kingdom, outbreaks still occur in Asia and Africa.

There are many other toxins produced by other fungi infesting mouldy cereals, fruit and vegetables. Any item of food which appears to be even slightly mouldy should be destroyed.

Farm chemicals

Farmers use many chemicals to protect crops and animals during production and to save harvested food from deterioration. The chemicals are used against weeds, insects, micro-organisms, rodents and external and internal parasites infesting farm animals. Without the use of farm chemicals food would be scarcer, more expensive and in some instances less safe. For these various reasons it is unrealistic to suppose that the use of all farm chemicals will cease in the foreseeable future. Thus the problem becomes one of adequate control of the use of these materials, the production of safe or safer ones and the stringent monitoring of food to ensure that farm chemical residues are well below toxic levels. In an industry as fragmented as farming and given that so much food is imported, control of farm chemical residues is not an easy task.

All food should be well-washed if it is not peeled. Items such as cabbages and lettuces can have their outer leaves discarded. Fruits such as apples and pears may have been sprayed with a wax-like substance which will not wash off, hence all apples and pears should be peeled before use. Oranges are also usually wax-sprayed and their peel is therefore not suitable for making marmalade. Unsprayed or organically grown orange peel should be used for preserves.

Organically grown food

The popularity of organically grown food is growing. As with all food, it is essential to discard any organically grown food which appears to be mouldy or diseased or in any way doubtful.

Chapter 16

Avoiding food-borne illness

Damage to health by food can be caused by the ingestion of micro-organisms in the food, by the ingestion of toxins produced by the organisms in the food, or by a combination of these. In addition, infestation by parasites may occur.

When enough toxin is ingested it will produce illness fairly quickly, usually within an hour. The most common effects are abdominal discomfort, nausea and vomiting, but often not diarrhoea. Other effects occur more rarely, usually affecting the nervous system and may sometimes be fatal (such as botulism). Cooking the food may destroy the toxin but this does not always happen and should not be relied upon. Food contaminated by toxin may appear to be safe, showing no change in colour or consistency and no abnormal smell or taste.

In contrast to food containing only toxin, food contaminated by dangerous micro-organisms causes illness only after the organisms have had time to multiply in the body, which may take hours or even days. Although the most common effects are abdominal pain, nausea, vomiting and diarrhoea, many other forms of illness may occur and may even be fatal. Food containing such organisms may taste or smell unusual but often it appears normal. Cooking the food so that its temperature is raised throughout to at least 70°C (158°F) for several minutes will almost always kill the organisms and will make the food safe provided little or no heat-stable toxin is also in the food.

Any food which may be unsafe should not be fed to pets.

In the shops

When buying food, choose a supplier whose premises are clean and in which the perishable food is kept cold or frozen in cooling units which are working properly as shown by a thermometer or recording device. These units should not be over-filled. The food packaging should be clean, intact and should display a clear 'use by' or 'best before' date. The staff should be clean and competent. When unpackaged food is being sold, the staff should not handle the food with bare hands but use disposable gloves. Cooked food should not be sold at the same counter as uncooked food. When buying cooked food which is going to be eaten without further preparation be very critical about the food display and the behaviour of the staff. Staff serving unpackaged cooked food should not handle money.

After purchasing perishable food, keep it as cool as possible until it is eaten, cooked or stored in a refrigerator or freezer. Do not allow it to remain in a warm car for an extended period.

In the kitchen

Before preparing food at home, always wash the hands with soap in running warm water. Dry the hands on a clean hand-towel (not the dish-drying towel) or with paper towels. If it is necessary to use a handkerchief or go to the toilet, wash the hands again before continuing food preparation. Keep the finger nails short.

When preparing food, do not touch pets, which should not be in the kitchen and should never be allowed on to surfaces used for food. Their feet and body surfaces are covered with micro-organisms. Any surface they sit on will be freely contaminated by their anal region. Keep kitchens free of flies.

Keep cooked and uncooked food well away from each other. Use separate utensils for cooked and uncooked food.

Do not lick the fingers. Use a clean spoon for tasting items (not the mixing spoon) and wash it before re-use.

Do not prepare food for others if the hands are infected.

Make use of spoons, forks and tongs rather than the hands for preparing food. Pick up utensils by the handle end. Use disposable gloves, especially when handling raw meat and poultry.

Food prepared some hours in advance should be kept in a refrigerator. Organisms will grow rapidly at kitchen temperature, often fast enough in one hour to cause illness.

All food should be carefully washed or peeled before cooking or serving raw.

Wash all crockery and utensils in hot water with detergent and then rinse well in warm running water.

> Frozen items are best thawed by placing in the refrigerator overnight.

Left-over food to be used again should be refrigerated or frozen as soon as possible and not allowed to stand in a warm room. Re-heating later may kill organisms which have grown before cooling but it may not always destroy any toxins released into the food.

Wash and dry all working surfaces thoroughly when food preparation is finished.

The most satisfactory way to thaw large frozen items is to place them in a refrigerator overnight so that they remain cool on the surface during thawing. If large items taking several hours to thaw are allowed to stand in a warm room there is the chance that organisms contaminating the outer parts may have ample time to multiply and perhaps produce heat-stable toxins. Do not refreeze frozen food once thawed.

In the refrigerator and freezer

Do not overload the refrigerator or freezer and do not put hot food into either. Allow spaces between stored items so that the air can circulate freely. Use a thermometer in the refrigerator to make sure the temperature is 1–4°C in the warmest place. Listeria grows well at 6°C but only slowly below 4°C. For the freezer, the thermometer must read −18°C or lower. Keep both pieces of equipment clean inside by carefully wrapping or sealing everything put into them. It is very important to prevent raw food from contaminating food which is to be eaten without further cooking.

On a picnic

Keep all perishable food in a cool-box with plentiful ice. Use tinned or bottled food if possible and open the containers just before eating. Take water, soap and towels for washing the hands before preparing the food and eating.

The 'at-risk' groups

These are the very young, the very old, all pregnant women and anyone with a deficient immune system. They should avoid soft cheeses, pâtés and pre-cooked meat, poultry, fish and shellfish to be eaten without further cooking. Hard cheeses, processed cheeses, cottage cheeses and cheese spreads are almost always safe. Salad items must always be carefully washed because they are sometimes contaminated by listeria organisms. All egg dishes must be cooked long enough to make the yolk solid. Items containing raw or only lightly cooked eggs should be avoided. All meat, poultry and fish must be thoroughly cooked so that no parts are even pink. Wrapped ice-cream is safer than scoop ice-cream.

Chapter 17

Exercise

Endurance

The ability to endure exercise depends partly on the amount of glycogen in the active muscles before the exercise begins, even though most of the energy for exercise is derived from the metabolism of fat. In some people, a diet rich in carbohydrate (more than 80 per cent of energy intake) prior to prolonged exercise can load up the muscles with up to twice the amount of glycogen found with a more balanced diet and this can be further increased by a more complex regimen of exercise and diet. For ordinary bursts of exercise, however, the normal content of glycogen in muscles suffices.

There is a prevalent belief that eating much protein will increase the ability to exercise. This is not so if the diet already contains adequate protein. During prolonged exercise a small amount of energy comes from protein, more comes from carbohydrate, but most comes from fatty acids brought to the muscles by the blood. Eventually, almost all the energy comes from these fatty acids. Eating extra protein, even just before exercise, in no way alters the nature of the metabolism of the exercising muscles. The belief that extra meat will improve performance may, of course, result in a sense of well-being after such a meal and that may enable an extra effort to be made. A vegetarian consuming an adequate diet may do equally well.

Vitamins

Any increase in metabolism increases the need for the vitamins B_1, B_2 and niacin. This is generally met by the vitamin content of a good mixed diet and of the extra food eaten to satisfy the increased energy output. However, if there is loss of body weight because food intake is insufficient foods particularly rich in these three vitamins should be eaten.

Water

An increase in water intake is essential for prolonged exercise, especially in warm conditions. If possible, small amounts of water should be taken during the exercise, but failing this enough water must be taken before the exercise begins. In the absence of sufficient water there is a fall in blood volume as much body water is lost via the skin and lungs, resulting in a poor delivery of oxygen and nutrients to the active muscles and a poor removal of metabolic products. The kidneys, also, will function less well.

The loss of water during prolonged exercise can be considerable, up to two litres when the temperature is moderate, with the loss rising to as much as four litres at high temperatures. During recovery, it may take an hour or more after drinking water for the adverse effects of water loss to wear off and the sense of thirst is not a good guide to how much water is needed. It is unlikely that too much water will be taken under these conditions.

For exercise of up to about one hour, plain cool water is probably as good as anything for rehydration. After one hour, some athletes find water mildly sweetened with glucose better than plain water, but the extra energy provided by the glucose is only small. Adding more than about 2.5 g of glucose to 100 ml of water will probably delay absorption of the water; if a strong glucose solution is used water will actually be lost from the blood to dilute the glucose in the stomach before water absorption can take place.

Salt

It is very unlikely that sweating during prolonged exercise will cause a significant loss of salt if a normal diet has been eaten. If there is any doubt about the body's salt balance the best way to deal with this is to add some extra salt to the usual food. The use of salt tablets during prolonged exercise can be harmful and will almost certainly delay the absorption of any water taken to combat dehydration. The various mixtures of salts in 'sports drinks' are probably of little significance in combating the fatigue accompanying prolonged exercise.

Exercise anaemia

In strenuously trained athletes, particularly runners, the blood haemoglobin is often below the average but as this is usually due to an increased plasma volume the total oxygen carrying power of the blood is not diminished. These athletes sometimes have free haemoglobin in their plasma as a result of red blood cell breakage in the pounding feet and violently contracting muscles. This will produce small amounts of free haemoglobin in the urine. There may also be small amounts of occult blood in the faeces. Sometimes very prolonged exercise, such as running a marathon, may result in appreciable blood loss in the faeces together with abdominal pain. Provided the diet is good there is rarely need for iron supplementation.

Chapter 18

Protein

Proteins, of which there are a vast number, have been known to be essential items of the diet since the early part of the nineteenth century, when they were found in animals and plants. They consist of long chains of relatively small molecules called amino acids. During digestion the proteins are split to release single amino acids which are carried by the blood to the liver and then to all the other cells of the body, where they are joined together in a new sequence to produce new proteins needed by the body. The amino acid sequence in a protein chain is of critical importance and, even though it may be made of hundreds of amino acids, if only one amino acid is in the wrong place the protein may no longer be suitable for its particular activity, or may perform that activity much less well.

There are about 12 kg of protein in the body of a normal 65 kg man and about 10 kg in a 65 kg woman. Approximately half is in the muscles, about one-fifth in the bones, one-tenth in the skin and the remaining one-fifth in the other tissues. Only the urine and the bile are normally without protein.

The body cannot store protein. Eating a lot of protein even over a prolonged period increases the adult body protein by less than 0.5 kg unless there is growth of muscle because of strenuous exercise. If the protein intake is more than is needed, the excess protein is either used for energy or is converted to fat, which is then stored. On the other hand, when the protein intake is less than is needed, body protein is rapidly lost. A healthy adult can lose 2–3 kg of protein before there is marked loss of function, the protein coming mainly from the skeletal muscles and the skin. As the tissues are in general about one-fifth protein, a loss of 3 kg of protein means that around 15 kg of body weight will be lost.

There is a loss of protein each day in the faeces, in rubbed-off skin, in lost secretions and in shed hair. This protein has to be replaced. In addition, the proteins of the tissues are not static but are constantly being broken-down and reformed. This turnover of tissue proteins is not 100 per cent efficient and some of the amino acids are metabolized and their nitrogen is lost in

> Protein is lost each day in faeces, rubbed-off skin, lost secretions and shed hair.

the urine, mainly in the form of urea. The small intestine is an extreme example of rapid protein turnover, its entire lining being shed and replaced every two days. The total weight of these cells is about 450 g, containing about 90 g of protein. Most of the amino acids in the protein of the shed cells are absorbed into the blood, but some loss in the faeces does occur. In contrast with the lining of the small intestine is bone, which has an extremely slow protein turnover rate. In an adult, about 250 g of body protein is renewed each day but as most of the old protein is re-used only about 40–50 g of dietary protein are required to balance this turnover.

Essential amino acids

Of the twenty-four different amino acids in body proteins, healthy adult cells can make sixteen for themselves but the other eight have to be provided ready-made in the dietary protein. These eight are therefore called essential amino acids. Their names are valine, leucine, isoleucine, lysine, methionine, phenylalanine, tyrosine and tryptophan. During periods of growth or illness some of the amino acid histidine is also needed ready-made. Dietary proteins containing a lot of these essential amino acids are good quality proteins in contrast to poor quality dietary proteins which are deficient in one or more of the essential amino acids.

It is unwise to consume pure single amino acids. Any pure single amino acid fed at a high dose will produce nausea, abdominal pain and vomiting. In the young there may be failure to grow normally because the high concentration of the excess single amino acid in the blood interferes with the proper use of the other amino acids. The use of amino acid supplements for body-building is very unlikely to give any advantage over eating good quality protein and may prove dangerous.

The single amino acid tryptophan is sometimes used to treat severe depression but this treatment must always be under the supervision of a suitable specialist.

Protein quality

A protein of high quality is one which will sustain health with rapid growth in the young and health with good tissue maintenace in the adult. To do this, about one-third of the protein must be essential amino acids. In the ordinary mixed diet no one protein is taken solely, the overall protein intake being a mixture of many proteins from animals and plants.

There are many ways of assessing protein quality, each measuring different but important effects on young and mature animals. One of these measures the amount of a protein that can be retained in the body during several days of eating that protein. A high quality protein will be largely retained in the body, whereas a poor quality protein will be metabolized for energy or fat production as it cannot be used for body protein because at least one essential amino acid is missing or is present in inadequate amount. This usage is called the Net Protein

Table 4 Protein quality of common foods

	Protein source	Quality measured as Net Protein Utilization estimated for children
Good quality protein	Human milk	94
	Hen's egg	87
	Cows' milk	81
Poor quality protein	Soya	67
	Rice	63
	Wheat	49
	Maize	36

For a mixed diet, the total protein quality should be 70 or more.

Utilization (NPU) and some values are given in Table 4, where it can be seen that human milk is exceptionally well utilized by children, whereas maize (sweet-corn) is of poor quality. For a mixed diet, the overall NPU should be above 70.

A simpler technique, good enough for ordinary dietary purposes, is to measure the weight gained by young rats for each gram of protein in their diet: it is called the Protein Efficiency Ratio (PER) and is used for labelling food in the United States and Canada. Experiments on animals, usually young rats, are very useful as guides for human nutrition but the results need to be confirmed in humans because discrepancies sometimes occur. For example, maize has a NPU value of 52 for rats but only 36 for children.

With any of the tests used, the best dietary protein for the human is of mammalian origin, the second best comes from poultry and the third from fish. Plant protein is generally of poor quality although by judicious mixing can be as good as mammalian protein (this is described under the section on complementary proteins below).

In an ordinary western mixed diet almost all the nitrogen in the food is in protein so that measuring the nitrogen content, which is easy, gives a good indication of the protein content. In vegetarian diets, however, much of the nitrogen is in non-protein form and often it cannot be used for protein synthesis by humans, so that the nitrogen in such diets can give misleading values for protein content. For animal foods, multiplying the grams of nitrogen by 6.25 gives a good indication of protein content. For western cereals, grams of nitrogen multiplied by 5.75 is used. These estimates need to be used with caution, especially for plant foods.

Complementary proteins

Poor quality proteins are poor for human nutrition because they have an inadequate amount of one or more of the essential amino acids. Thus pulses (called

On a vegetarian diet always mix pulses with cereals at each meal.

legumes in the United States) have poor quality protein because there is a deficiency of the essential amino acid methionine, but they are rich in lysine. Young animals fed on only pulses fail to grow well. They also fail to grow well if fed only cereals, which are poor in lysine but are rich in methionine. Young animals thrive, however, if fed a mixture of pulses plus cereals because the deficiency of methionine in the pulses is made good by the rich supply of methionine in the cereals; and the lack of lysine in the cereals is made good by the rich lysine content of the pulses. They are therefore known as complementary proteins. Hence, although bread by itself is not a source of high quality protein and nor are baked beans by themselves, a meal of baked beans (pulse) plus bread (cereal) gives protein as useful as meat protein and is very inexpensive. The cereal maize (sweetcorn) does not fit into this scheme because it is deficient in the essential amino acid tryptophan.

A meal of baked beans on toast gives protein as useful as meat protein.

Many poor quality proteins are adequate for general tissue maintenance even though they may not sustain growth or tissue healing or be adequate during severe illness. They can, however, be greatly improved by the addition of small amounts of the good quality protein in meat, poultry, milk, cheese and egg.

The common cereals in the United Kingdom are wheat, rice, maize, oats, rye, soybean and barley. Pulses are all kinds of beans (except soybean, which is a cereal), peas and lentils.

Protein requirements

A minimum quantity of protein is required in the diet throughout life in order to provide the essential amino acids. During periods of rapid growth, as in infants and young children, the need for good quality protein is great, whereas in healthy adult men the need is at its lowest. Other groups who need more than a maintenance amount of protein are pregnant women, nursing mothers, severely ill people, anybody in a state of severe anxiety or stress and post-operative patients. For all these, good quality protein is necessary.

The minimum amount of protein needed for good health has been estimated for different groups of subjects. The simplest estimate is to measure the amount of nitrogen lost in the urine and the faeces when a diet without protein is fed for several days. On average, healthy adult men weighing 70 kg lose about 7 g of nitrogen each

The body does not use high quality protein with 100 per cent efficiency so extra has to be eaten to make up for this wastage.

day on such a diet, which is equivalent to about 45 g of protein per day. There is a big variation within this group, the lowest loss per man per day being as little as 20 g of protein and the highest as much as 65 g. It would seem, therefore, that 65 g of protein would be enough for all healthy adult men. However, the body does not use even good quality protein with 100 per cent efficiency, so that extra has to be eaten to make up for this wastage, which is much greater in some people than in others. In addition, the more protein eaten, the greater the wastage becomes. To allow for these variations, recommended protein intakes usually include a safety margin of an extra 25–30 per cent above the value that would seem to be adequate for about 98 per cent of the group under consideration. If the dietary protein is of poor quality, the daily intake must be increased. This is especially important for those whose protein needs are in excess of just maintenance levels. There is no need for extra protein during periods of strenuous work unless muscles are getting bigger.

For children and still-growing young people, the protein needed each day for normal growth and development can be assessed. This requires the measurement of the increase in body protein which, because it is time-consuming and therefore expensive, is rarely done.

When there is not enough carbohydrate and fat in the diet to supply the energy needed, some of the protein is used and is hence unavailable for tissue maintenance or growth. On a low energy intake, as during slimming, the food should therefore be rich in good quality protein.

> Protein in a vegetarian diet is only 85 per cent digested compared with 98 per cent or more for animal protein.

Extra protein is needed when it is poorly digested. For example, protein in a vegetarian diet is only about 85 per cent digested compared with 98 per cent or even more for animal protein. Digestion can also be reduced in intestinal disease because of inadequate digestive enzymes. In severe diarrhoea the food may move along the intestine so fast that there is not enough time for complete digestion.

Each day, about 70–80 g of good quality protein are needed by healthy adult men and non-pregnant, non-lactating healthy adult women consuming about 2500 kcal. One-third of this is normally of animal origin and the rest comes mainly from cereals and pulses. Vegetarians can use milk, milk products and eggs; for vegans, mixed cereals and pulses can provide an adequate supply of essential amino acids. In most parts of the world the average daily protein intake is 50–100 g. This is generally enough except where the quality of the protein is especially poor or where there are diseases requiring a particularly high protein intake. In the western world the daily protein intake is on average 75–100 g and as the protein quality is

> In the United Kingdom there is virtually no dietary deficiency of protein except in disease or severe alcoholism.

Table 5 Protein content of common foods

High (More than 20 g/100 g edible portion)	Beef (lean), cheese (hard: Caerphilly, Cheddar, cheshire, double gloucester, edam, Emmental, Gorgonzola, gouda, Gruyère, lancashire, red leicester, stilton, wensleydale), chicken (no skin), duck (no skin), haddock, ham (lean), heart, kidney, lamb (lean), liver, peanuts, rabbit, tongue, turkey (no skin), veal
Medium (10–20 g/100 g edible portion)	Almonds, beans, beef (fat), Brazil nuts, cheese (soft: brie, Camembert, danish blue, lymeswold, processed), cod, egg, hazelnuts, herring, lamb (fat), mackerel, pork
Low (2–10 g/100 g edible portion)	Avocados, bread, Brussels sprouts, cabbage, chocolate, cornflakes, figs (dried), lentils, maize, peas, potatoes, prunes, rice, spinach, watercress
Very low (Less than 2 g/100 g edible portion)	Apples, apricots (fresh), butter, carrots, cauliflower, celery, dates (dried), fats, grapefruit, honey, lettuce, mushrooms, oranges, pears, plums, sweet peppers, tomatoes

high this intake is more than needed. There are parts of the world where protein intake is sufficiently inadequate either in amount or quality as to cause disease. In the United Kingdom there is virtually no dietary deficiency of protein except in disease or severe alcoholism.

About one-fifth of lean uncooked meat, poultry and fish is protein, so that 100 g of these foods provides about 20 g of high quality protein. This with one egg (about 6 g of protein), milk and milk products will give at least one-third of a healthy adult's daily protein need. The rest of the day's protein will be in bread, other cereals, peas, beans and lentils. Although nuts are a rich source of protein, usually too few are eaten in the United Kingdom to be of dietary significance.

If a family has only limited dietary protein, the young, the pregnant and the lactating women must be given an adequate supply of the protein and the healthy adult men what remains as their need for protein is the least.

The needs of pregnant women, nursing mothers, children and other special groups are discussed in the appropriate Chapters.

The protein content of common foods is given in Table 5.

High-protein diets

A diet containing two or three times the needed amount of protein is harmless and many groups of people regularly consume up to 200 g of protein per day. If, however, very large amounts of protein are suddenly introduced into the diet, the great rise in urea production, the way in which unneeded nitrogen from the protein is excreted by the kidneys, may cause severe dehydration as the volume of urine becomes excessive. This can happen when formula milk is not prepared properly and too much powder is used, causing severe infant dehydration. In adults, the body can adapt to a very high protein intake if the increase is slow. It is possible to consume up to 75 per cent of energy need as protein, which amounts to about 450 g of protein (about 5 lb of raw meat) for a daily intake of 2400 kcal.

> A diet very high in protein conveys no advantage over a normal diet in healthy people.

Diets containing more protein than is needed do not confer any health advantage. Animals fed just the amount they need can perform exercise and overcome unfavourable temperatures, injuries and infections equally well as animals fed high-protein diets. The same seems to be true for humans, many of whom remain just as healthy on 50 g of protein per day as do others eating over 200 g of protein per day.

Protein deficiency in childhood

In the absence of disease, protein deficiency in childhood in the United Kingdom is very rare, occurring mostly as a result of parental food fads. When it does occur, it starts soon after weaning. The child looks unwell and does not grow at the expected rate; it is usually pale, flabby and listless. As the condition worsens the dependent parts of the body swell due to accumulation of fluid in the tissues (oedema); there is often a pot-belly produced by oedema in the abdomen and an enlarging fatty liver; the arms and legs are very thin. When there is an adequate energy intake and the protein deficiency is not too extreme the child survives but is physically stunted and mentally retarded. This condition bears the name 'kwashoirkor'.

If there is an inadequate energy supply as well as lack of protein, the outcome is grave because some of the dietary protein, already too low, is used for energy and is hence not available for tissue growth and development. This condition is known as 'marasmic kwashiorkor'. The child hardly grows but the considerable oedema that develops may make the lack of growth less apparent. Weighing such a child gives no useful information about growth because of all the water in the tissues.

The treatment for these children is to feed an ample amount of good quality protein with enough carbohydrate and fat to satisfy the energy need. In addition, vitamin supplements must be given. As the synthesis of new protein progresses there is a rise in the level of the proteins in the blood and this cures the oedema, the excess fluid being removed from the tissues and excreted in the urine, so that there is a fall in body weight. This is a good sign and shows that the child is recovering. If treatment is started before the age of 4–5 years and the protein deficiency not too extreme, the child may eventually become normal or almost so. If, however, treatment is delayed or the original deficiency great the child will remain stunted and mentally retarded.

Damage to protein during food preparation

Boiling milk and spray-drying it to produce dried milk powder does virtually no damage to the protein. The process of making evaporated milk does destroy about 10 per cent of the essential amino acid lysine but this is very unlikely to be important. Storing dried milk powder at room temperature for several years may diminish the availability of some of the essential amino acids. The powder becomes darker over this time, which will indicate that storage has been too long. Similarly, evaporated milk stored at room temperature for several years may show browning.

Severe heating of food may make some proteins less useful by making them less digestible. Repeated cooking of meat usually destroys some of its essential amino acid methionine.

When protein is heated in the presence of some sugars there is a reaction resulting in the loss of some of the essential amino acids lysine and tryptophan. The food goes brown.

Chapter 19

Carbohydrate

The dietary carbohydrates come almost entirely from plants, although there is a very small amount in animal foods. They are conveniently divided into two groups, the simple carbohydrates and the complex carbohydrates.

The simple carbohydrates

The simple carbohydrates in the diet are of two sorts: those consisting of only one molecule (the monosaccharides) and those which are made of two molecules (the disaccharides).

The main single-molecule carbohydrates in food are fructose, glucose and mannose. Fructose is in almost all fruits and is abundant in honey, which is a mixture of fructose and glucose. It is the sweetening agent of antiquity. Glucose, present in many fruits, especially grapes, and in vegetables, is less abundant than fructose. Mannose is also in many fruits but usually in only small amounts.

The dietary two-molecule carbohydrates are sucrose, lactose and maltose. Sucrose is table sugar and consists of fructose plus glucose. Lactose is milk sugar and is glucose plus galactose (which is a single-molecule carbohydrate that occurs only in this combination). Maltose is a combination of two molecules of glucose and is found in malted barley. Pure sucrose is white. A less refined product, called raw sugar, contains some of the original constituents of sugar-cane and is brown; it has no significant nutritional advantages over pure white sugar and may contain undesirable substances. Brown sugar is sometimes made by colouring pure white sugar with caramel.

Fructose, glucose, mannose, sucrose, lactose and maltose are all known as 'sugars' in chemical terminology, which may be confusing because in every-day language sugar means table-sugar, which is sucrose. There are many other simple carbohydrates in fruits and vegetables but they are of little significance in the ordinary diet. For healthy people the use of pure glucose, which is expensive, offers no nutritional advantage over ordinary white table sugar, which is cheap.

The complex carbohydrates

The main complex carbohydrates in the diet are starch, the form in which energy is stored by plants, and cellulose, which gives plants their structure. There are also in plant foods hemicelluloses and pectins, while foods of animal origin may have small amounts of glycogen. These three materials are also complex carbohydrates.

Starch and cellulose, both made of many glucose molecules, occur in abundance in the ordinary diet. Only starch, however, can be digested by humans, being broken-down in the intestine to simple glucose molecules, which are then absorbed into the blood. The cellulose along with the hemicelluloses and pectins in the diet pass through the small intestine undigested and enter the large intestine (the colon) where they are either used by the micro-organisms which live there or are excreted intact in the faeces, giving bulk and softness.

During ripening and storage the sugar content of some plants is turned to starch, so that peas, maize, beans and carrots become less sweet. In contrast, some of the starch in apples, pears and bananas is turned to sucrose on ripening and storage so that these items become more sweet.

Glycogen

The complex carbohydrate glycogen in foods of animal origin is also made of many glucose molecules. It is found in liver and in muscle but it usually provides only a very small part of the daily energy requirement.

In the human the store of glycogen is quite small, amounting to around 450 g (1 lb), about 150 g being in the liver and the rest in the muscles. As the energy value of 1 g of carbohydrate is about 4 kcal, the store of glycogen under ordinary conditions yields about 1800 kcal, representing perhaps three-quarters of one day's energy need. This is trivial when compared with the normal store of 10–20 kg of fat with its energy yield of about 9 kcal per gram. It is possible to double the glycogen store of the muscles by eating a diet very high in carbohydrate and exercising vigorously between each meal, a technique used by athletes who require a high rate of energy expenditure over about half an hour. Despite this carbohydrate loading, almost all the energy in strenuous exercise comes from the metabolism of fatty acids.

It is possible to double the glycogen store of the muscles by eating a diet very high in carbohydrate and exercising vigorously between each meal, a technique used by athletes who require a high rate of energy expenditure over about half an hour.

The body store of glycogen is used to replenish the glucose level of the blood. When virtually all the glycogen has been used, as it will be on fasting for about 16 hours, further glucose can be produced from body protein. To prevent this conversion of body protein to glucose it is necessary for an average adult to eat at least 100 g of carbohydrate each day.

Heating plant food

Plant food should be cooked until tender.

Unlike animal cells, which are easily digested by enzymes in the human small intestine, plant cells have walls of cellulose and similar material which are not digested by human enzymes. To obtain the nutritionally important contents of plant cells, the cells must be disrupted. Grating and chewing can do this to only a limited extent, but heating plant food until it is tender opens the cells and allows the contents to become available. If plant food is eaten inadequately heated, much of it enters the large intestine undigested and its nutritional value is lost. Although heating food may occasionally destroy part of its nutritional value, this is rarely of importance for the ordinary UK diet.

Daily intake of carbohydrate

In the early part of infancy the only necessary carbohydrate is the disaccharide lactose, present in the mother's milk or infant formula. After about six months a gradually increasing quantity of properly prepared plant food should be introduced into the weaning diet. For many infants sucrose soon becomes an important source of energy but under ordinary conditions its use should be limited.

In the current ordinary UK adult diet starch provides about half of the daily need of about 2400 kcal. As 1 g of starch yields about 4 kcal, this represents about 300 g of starch (10–11 oz); the weight of the plant food containing this starch will vary greatly with the variety eaten. In poor countries, where less protein and fat are available, starch may provide over 80 per cent of daily energy need. In contrast, for communities such as the Inuit living in a traditional way, only about 10 per cent of their energy is derived from carbohydrate.

One of the greatest changes in diet has been the remarkable increase in sucrose (table sugar) consumption. In 1900 the world production of sucrose was about 8 million tons per year, yet by 1965 it was over 50 million tons per year. Until the eighteenth century sucrose was an expensive luxury in the United Kingdom but by the 1980s this single compound accounted for 15–20 per cent of the daily energy intake, being on average about 100 g (3.5 oz) of sucrose per person per day. Although sucrose is a very valuable addition to the diet, providing an attractive source of cheap energy and making many foods more palatable, its current excessive use needs to be curtailed. It is very likely to increase obesity and it is a major cause of damaged teeth, especially in the young. This is particularly so in the lower income groups, who consume about twice the sucrose of the more affluent.

Although there have been many claims that several important diseases result from the current sucrose consumption, a UK report in 1989 found no evidence in most normal people of a direct adverse effect on blood cholesterol levels, on the

Table 6 Carbohydrate content of common foods

High (More than 50 g/100 g edible portion)	Brazil nuts, bread, chocolate, cornflakes, dates (dried), figs (dried), honey, prunes, raisins
Medium (10–40 g/100 g edible portion)	Apples, bananas, beans, lentils, maize, pears, peas, potatoes
Low (1–10 g/100 g edible portion)	Almonds, apricots (fresh), avocados, Brussels sprouts, cabbage, carrots, cauliflower, celery, grapefruit, milk, oranges, peanuts, plums, spinach, sweet peppers, tomatoes, watercress
Very low (Less than 1 g/100 g edible portion)	Beef, butter, cheese, chicken (no skin), cod, duck (no skin), egg, fats, haddock, ham, heart, herring, kidney, lamb, lettuce, liver, mackerel, mushrooms, oils, pork, prawns, rabbit, salmon, tongue, turkey (no skin), veal

blood triglyceride level, on cardiovascular disease or on high blood pressure; nor was there evidence of a causal link between sucrose consumption and abnormal behaviour except for some extremely rare metabolic disorders.

Good sources of carbohydrate in an everyday UK diet are given in Table 6.

There are essential amino acids in protein and essential fatty acids in fat but there are no essential carbohydrates. Metabolizable carbohydrates are all equally useful. All that is required is that at least 100 g of carbohydrate (any sort) should be in each day's diet.

Relative sweetness

The simple carbohydrates commonly occurring in everyday UK food are sweet, although some are much sweeter than others. Starch, in contrast, is not sweet. There is no absolute measure of sweetness but sweet things can be compared with each other to give a relative sweetness value. If sucrose is given a sweetness value of 100, then fructose at 170 is nearly twice as sweet, while at 65 glucose is only two-thirds as sweet. Most of the sweetness of fruits comes from the fructose in them. Table 7 gives a list of relative sweetness for carbohydrates and in addition gives values for four sugar substitutes. Three of these substitutes are indeed very sweet, especially saccharin.

In ancient times the dietary sources of sweetness were fruits and honey, both often scarce and seasonal. Honey contains almost no protein and only trivial amounts of minerals and vitamins. Its only value is to provide energy from its carbohydrate content and to add flavour and sweetness. Although often greatly praised it has no known special ingredients to improve nutrition or health. Its approximate composition is given in Table 8.

Table 7 Relative sweetness

	Approximate sweetness
Saccharin	40,000
Aspartame	18,000
Cyclamate	3,000
Fructose	170
Sucrose	100
Glucose	65
Sorbitol	50
Maltose	45
Lactose	30

Table 8 Composition of honey

	Per cent (%)
Fructose	40
Glucose	30
Water	20
Other sugars	8
Other ingredients	2

Sucrose substitutes

Because the consumption of sucrose increases the energy intake and because sucrose is known to be an important cause of tooth decay, substitute sweeteners have been sought. Among the best known are saccharin, aspartame, cyclamate and sorbitol. Saccharin, which has no energy value, has been used for over 100 years. Cyclamate also has no energy value. Aspartame is made of two amino acids (aspartic acid plus phenylalanine) and has virtually no energy value in the amounts used as a sweetener. Because aspartame contains phenylalanine it must not be consumed by people who have the genetic disorder of metabolism called phenylketonuria (PKU). Sorbitol has been used for many years as a substitute for sucrose in diabetic foods. It is a naturally-occurring derivative of glucose and is found in small quantities in some fruits; it is, however, made commercially for use in the food industry. It adds both sweetness and bulk to food and because it is converted by the liver to fructose it has only a small effect on the blood glucose level, which is important for diabetics. Some ingested sorbitol escapes absorption by the small intestine and enters the large intestine, where it may cause flatulence and diarrhoea if too much is eaten.

One of the disadvantages of saccharin, aspartame and cyclamate is that very little is needed to add sweetness to the food when compared with sucrose. This means that another substance is sometimes needed to replace the bulk lost by not using sucrose. These artificial sweeteners do not cause tooth decay.

There have been claims that aspartame may cause visual damage, even blindness. The U.S. Food and Drugs Administration has been unable to confirm this and aspartame does not seem to be the cause of any visual damage.

Lactose intolerance

Lactose intolerance is the name given to the condition in which the milk-sugar lactose cannot be fully digested. It is not seen in the very young because the wall of their small intestine contains enough of the enzyme lactase, which digests the lactose to glucose plus galactose. As the child grows, however, some of the lactase gradually disappears. Often there is sufficient left to deal with small amounts of lactose but this may not be adequate for the amount of lactose encountered in milk-drinking western countries. If more than about 10 g of lactose is consumed at one meal (about 300 ml of milk [half a pint]), some of the lactose may enter the large intestine where micro-organisms metabolize it, sometimes causing severe abdominal discomfort and diarrhoea. The symptoms clear up rapidly if lactose is eliminated from the diet or kept very low. To avoid lactose the subject must cease using ordinary milk and products containing lactose. Fortunately, lactose is rarely used as a sweetener or for bulk. Cheese, although it is a milk product, may be eaten because it contains only negligible amounts of lactose.

> Lactose intolerance is not an allergy because the immune system is not involved.

Lactose intolerance is widespread throughout the non-white people of Asia and Africa, but is uncommon in white populations, who mostly continue to produce adequate amounts of lactase during their adult life. The problem is not usually important in Asia and Africa, where traditionally little milk is used by adults, but it may become very troublesome when people of Asian or African origin settle in white communities where much milk is used in the general diet. It is possible to get milk in which all the lactose has been split to glucose plus galactose by the addition of plant enzyme. In the United Kingdom about 5 per cent of the white population is partially lactase deficient but because intolerance is dose dependent only 1–2 per cent develop symptoms.

In addition to this common racial-ethnic form of lactose intolerance, there are other very uncommon forms of the condition.

Ketosis due to lack of carbohydrate

For the normal complete metabolism of fat by the body tissues, some carbohydrate in the diet is essential. People vary considerably in this respect, but the average minimum daily adult intake needed for the prevention of ketosis is about 100 g. Almost all palatable diets contain much more than this, usually about 300–500 g per day. In the absence of enough dietary carbohydrate there

Diets very low in carbohydrate can be harmful.

is only partial metabolism of fat, which results in an accumulation of two unwanted acidic substances, one being β-hydroxy-butyric acid and the other aceto-acetic acid. The blood becomes more acidic than is normal, producing at first only malaise, loss of energy and dehydration but a fatal outcome may follow if the condition is left untreated. In addition to the two excess acids, a third substance called acetone is produced to excess, giving the breath a characteristic smell.

Ketosis can occur not only when carbohydrate is lacking in the diet but also in starvation, when the metabolism is using the body fat. Some glucose can be formed from the body's protein and a few individuals can thereby avoid dangerous ketosis during starvation, but most people cannot.

Carbohydrate additives

Many complex carbohydrates are used as food additives. As most cannot be metabolized they are useful for low-calorie products. Agar and carrageenan, derived from algae, are used for bulk, gelling and as fat stabilizers. Pectic substances from citrus peel, and guar gum from a legume are also used as gelling agents. Methylcellulose, made synthetically, is employed for bulk in low-calorie foods. They all add to the dietary fibre intake, which is discussed in Chapter 23.

Hypoglycaemia

Hypoglycaemia is very rare.

An abnormally low level of glucose in the blood is called hypoglycaemia and, except in diabetics taking too much insulin, it is a rare condition. The blood glucose level is very carefully controlled and is kept adequate even during prolonged fasting and only shortly before death does it fall markedly. This careful control is needed because the brain normally uses only glucose for its energy supply. Almost all diagnoses of hypoglycaemia are likely to be wrong unless they have been checked by measuring the true glucose concentration in the blood.

The early symptoms of hypoglycaemia are hunger, trembling, sweating and unsteadiness, followed later by convulsions, coma and death. Some subjects may become confused and violent, as if drunk.

There is a belief that hypoglycaemia is common in young people and is a cause of delinquency. The removal of sucrose from the diet is said to be curative. There is no scientific evidence for this belief.

Chapter 20

Fat

Under natural conditions dietary fats are rare. Apart from nuts, avocados and olives, everyday fruits and vegetables have practically no fat and most wild animals are thin. From earliest times, fats have been sought after and have symbolized rich living: to kill the fatted calf; to live off the fat of the land. In many poor countries fat still provides only about 5 per cent of the energy intake, but modern agriculture together with new technology have changed the dietary scene drastically in the affluent countries so that about 40 per cent or more of the daily energy intake now comes from fat. Instead of being a food for special occasions, fat, in these richer countries, is the cheapest source of dietary energy and its excessive use has brought many health problems.

> The fat content of a food cannot be judged from its appearance.

Fat is present in the diet as visible fat, either on the surface of the food or marbling it, or it may be present as invisible fat within the substance of the food. This invisible fat may lie between the cells or be within the cells. Even lean-looking food may contain a considerable amount of fat. Apparently lean meat may be rich in fat, especially if it has been processed in some way because it is possible to disperse fat within meat so that it cannot be seen.

The energy value of fat is high, yielding about 9 kcal/g, in contrast to the 4 kcal/g for carbohydrate and protein. In addition, foods rich in fat usually have little water, so that much can be eaten before a sense of fullness occurs.

For many people, though not often children, fat in the diet makes it more palatable. This may be partly because low-fat foods are poorly presented by a food industry engaged in producing high-fat food.

Types of fat

Dietary fat (sometimes called lipid) is made of substances called glycerides. These are composed of fatty acid molecules joined to glycerol (glycerine). Most of the dietary fatty acids have more than twelve carbon atoms in each molecule and are called long-chain fatty acids; there are others with fewer carbon atoms and they are called short-chain fatty acids. Apart from butter almost all glycerides in a normal UK diet are made of long-chain fatty acids.

The fatty acids of dietary fat are of two sorts: saturated or unsaturated. This describes the type of bonds (joins) between the carbon atoms of the fatty acid molecule. In saturated fatty acids the bonds are all of the single type, while in unsaturated fatty acids there is at least one double bond. When there is only one double bond the fatty acid is called mono-unsaturated, while with two or more double bonds the fatty acid is said to be polyunsaturated. At room temperature (about 18°C or 64°F) fats made mainly of saturated fatty acids are solid (like butter), whereas fats containing mainly unsaturated fatty acids are liquid (they are oils). Dietary fat, therefore, may consist of saturated fatty acids, mono-unsaturated fatty acids and polyunsaturated fatty acids.

For nutritional purposes it is useful to know how much polyunsaturated fat and how much saturated fat there is in food. This is called the P/S ratio. With a high P/S ratio the fat at room temperature will be soft or even liquid; with a low P/S ratio the fat will be hard. Generally, animal fats have a low P/S ratio and plant fats have a high P/S ratio. The fats of poultry and fish have a higher P/S ratio than the fats of cows, pigs and sheep.

It is possible to convert unsaturated fat to saturated fat by a process called hydrogenation. Because plant fats are cheaper than animal fats this is often done in food manufacture when saturated fat is wanted. One drawback of hydrogenation is that it destroys much or all of the essential linoleic and linolenic polyunsaturated fatty acids.

Omega-3 fish oils

These are long-chain polyunsaturated fatty acids found in fatty fish. In some people these oils lower the undesirable low-density-lipoprotein cholesterol and the triglyceride levels in the blood. About 100 g of fatty fish probably provides enough of these oils. Some shops sell bread incorporating omega–3 fatty acids.

Essential fatty acids

> Linoleic acid and linolenic acid must be present in food because the body cannot synthesize them. They are essential fatty acids.

Although normal dietary fats contain many different fatty acids, only linoleic acid and probably linolenic acid have to be present in the food because human cells cannot synthesize them. They are therefore referred to as essential fatty acids. A third acid, arachidonic, is a borderline essential fatty acid because it can be made by the cells when there is a rich supply of linoleic acid.

The daily requirement for linoleic acid for an adult is not known accurately and is given as 2–8 g per day; most diets in western countries provide at least twice this amount. It is extremely rare to find adults suffering from essential fatty acid deficiency. The few who have been described have been fed for several weeks on fat-free diets because of disease. Children are more susceptible and have developed signs of essential fatty acid deficiency after about one week on a fat-free diet. These cases occurred before the need for dietary linoleic acid was recognized.

Linoleic and linolenic acids are plentiful in stored body fat and during weight loss they are released for general body use as the stored fat is metabolized. During dieting, therefore, even though little fat may be eaten, there will not be a deficiency of essential fatty acids. At least 500 g of linoleic acid are in the fat stored by an adult.

Linoleic and linolenic acids are essential for normal growth of all cells and together with arachidonic acid are needed for the production of a collection of hormone-like substances called prostaglandins, which are involved in a large number of very important reactions in the normal and abnormal or diseased states. In addition, diets high in essential fatty acids lower the blood cholesterol level.

Apart from these essential fatty acids, the rest of the dietary fat is not essential, provided that sufficient energy is supplied by carbohydrate and protein. The vitamins which are fat-soluble (A, D, E and K) can be obtained without eating fat; this is described in Chapters 26–35.

Daily dietary fat

Very poor countries tend to have diets containing very little fat, in some cases as little as 5 g per day, about 2 per cent of the energy intake, and although the populations often have general ill-health they do not seem to be in need of fat. Virtually all their fat comes from plants, providing them with the necessary 2–3 g of essential fatty acids per day. Diets in affluent countries, however, are loaded with fat (up to 150 g per day), mostly saturated, giving up to 40 per cent of the daily energy intake. Much of this fat is of animal origin or is partially hydrogenated plant fat. Its essential fatty acid content is nevertheless satisfactory.

The maximum amount of fat that can be metabolized normally by adults is about 2.5 g, fat/kg body weight/day, which is around 175 g, fat for an average 70 kg person, yielding about 1575 kcal per day, which is more than 50 per cent of the average daily energy requirement. The current consumption of fat in most western countries, about 40 per cent or more of the daily energy need, is therefore approaching the maximum tolerable fat intake. Sources of dietary fat are given in Table 9.

The dietary fats rich in polyunsaturated fatty acids are the ones likely to be rich in essential fatty acids. Very good sources of essential fatty acids are safflower, maize, cottonseed and peanut oils. Olive oil, butter and egg yolk are poor sources. Some margarines and fat spreads are very rich in essential fatty acids. Poultry fat is also a good source as are the oils of walnuts and Brazil nuts. The

Table 9 Fat content of common foods

Very high (More than 75 g/100 g edible portion)	Butter, dripping, fat spreads, lard, margarine, oils
High (25–75 g/100 g edible portion)	Almonds, Brazil nuts, cheese (hard), chocolate, hazelnuts, peanuts, walnuts
Medium (10–25 g/100 g edible portion)	Avocados, beef (medium fat), egg, herring, lamb (medium fat), mackerel, pork (medium fat), salmon, tongue
Low (2–10 g/100 g edible portion)	Beef (lean), chicken (no skin), duck (not skin), ham (lean), heart, kidney, lamb (lean), liver, pork (lean), rabbit, soybean, turkey (no skin), veal
Very low (Less than 2 g/100 g edible portion)	Beans (except soybean), bread, cod, fruit (except avocados), haddock, honey, prawns, rice, vegetables

amount and type of fat in meat and poultry can vary considerably from sample to sample depending on the food fed to the animals, whereas the fat in fish and plants is much less variable.

The percentage of the daily energy derived from animal fat can be very high. The 3 per cent or so of fat in whole milk may not seem much but it yields about half the total energy supplied by the milk. For products such as minces, sausages, pâtés and meat pies about three-quarters of their total energy yield may come from animal fat.

Ice-cream

Ice-cream does not have to contain cream but if it is called dairy ice-cream all the fat must come from milk. Non-dairy ice-cream is usually made of only vegetable fat, often coconut and palm oils, which are mainly saturated fats, as are the fats in milk. The texture of ice-cream depends not only on its solid ingredients but also on the air it contains. If ice-cream is allowed to thaw so that the air escapes, refreezing the liquid ingredients produces a very different product. Non-dairy ice-cream does not always taste worse than dairy ice-cream, is not always cheaper and does not always have less energy content than dairy ice-cream. Nutritionally there is often little difference between dairy and non-dairy ice-cream.

Olestra (Proctor and Gamble)

This is a relatively new fat substitute made by joining fatty acids to sucrose, hence it is not a glyceride or true fat. It is said to have the taste and cooking properties of a true fat and is currently used mainly in snack foods. It is not digested or absorbed into the blood by humans and hence provides no energy, which is the reason for its use. In some people it has caused flatulence, soft stools and occasional faecal leakage. It is fortified with vitamins A, D, E and K to prevent it leaching out these substances from the diet and their loss in the faeces.

Fat-controlled diets

The need to reduce the total fat intake of most people in the richer countries has been emphasized repeatedly by many medical organizations. In general, the energy derived from dietary fat should be in the range 30–35 per cent of the total daily energy intake instead of the current 40 per cent. The energy from saturated fat should be about 10 per cent of the total energy intake and that derived from polyunsaturated fatty acids should be about the same, leaving about 10–15 per cent to come from mono-unsaturated fat. The polyunsaturated to saturated fat ratio should be about 1.0. The reasons for these recommendations are that the blood levels of total cholesterol and the undesirable low-density-lipoprotein cholesterol can be considerably influenced in some people, called responders, by their saturated fat intake. In these people, lowering the saturated fat in the diet reduces the cholesterol levels, which reduces the risk of coronary heart disease. Reducing the cholesterol intake has little effect on the blood cholesterol levels. For non-responders, the small benefit seen with dietary change might not be worth altering food habits and if these people need to lower their cholesterol levels they will have to take one of the several useful medicines now avaialble.

> Reduce fat in the diet, especially saturated fat.

Because it is not feasible to test everybody to discover whether or not their blood cholesterol levels respond to a low saturated-fat diet and because a low fat intake is generally desirable not only to diminish the risk of coronary heart disease but to reduce the incidence of obesity and its complications most people would do well to keep to the following guidelines:

a) Limit the use of beef, lamb, pork, ham, minces, pâtés, sausages, meat pies, offal, lard, butter, full-fat cheese.
b) Increase the use of fish, poultry (without the skin), skimmed or semi-skimmed milk, low-fat yogurt, low-fat cheese.
c) Trim off all visible fat.
d) Avoid fried foods.
e) Avoid most snack foods (choose those with the least fat).
f) Cook with a minimum use of fat; use vegetable and nut oils high in mono-unsaturated and polyunsaturated fat.
g) Avoid chocolate, cakes, biscuits, high-fat ice-cream.
h) Use more vegetables, potatoes (boiled or baked in jackets), wholegrain rice, wholegrain pasta, salads, fruit.

There are difficulties in giving more than general advice to populations as a whole because of the marked variations in local diets and the possible genetic differences between groups of people. For example, the Greeks currently have a very high fat intake, amounting to about 43 per cent of their total energy consumption, yet they have a very low rate of coronary heart disease. Advising them to reduce their fat intake to avoid coronary heart disease would be foolish. The reason for their low risk of heart disease is not known but it may be due to the protective action of the mono-unsaturated oleic acid in their high olive oil consumption, or it may be genetically determined, or some other factors may be involved.

Infants and young children should not be fed a low-fat diet because it may cause them to become energy-deficient. By the age of 5–10 years, however, it is desirable to keep the fat in the diet down and to ensure a good intake of polyunsaturated fat so as to prevent the beginning of the process that will lead to adult coronary heart disease. Although the chemical composition of the blood is probably to a large extent under genetic control good nutritional habits in the early years of life seem also to play an important role in determining adult disease.

Cancer and polyunsaturated fatty acid

There is a theoretical risk that consuming large quantities of polyunsaturated fatty acid might cause tumours. This is because polyunsaturated fatty acid is prone to oxidation, generating substances called free radicals which can damage cell nucleic acids. Experiments with animals fed about 5 per cent of their energy intake as polyunsaturated fat have shown increased tumour incidence. This level of polyunsaturated fat is about the same as that currently recommended for human consumption. In contrast with these animal experiments, however, studies in people have not shown evidence suggesting that eating polyunsaturated fat is associated with cancer. Nevertheless, in view of the uncertainty of the effect over many years, it seems prudent not to exceed about 10 per cent of energy intake as polyunsaturated fat.

Cooking in fat

Heated fats for cooking (frying, roasting) appear to be harmless when consumed at ordinary dietary levels, even over long periods, except in so far as they add to the fat intake. Good practice is to use oil with a high polyunsaturated fatty acid content, not to heat the oil above 200°C and to renew the oil after 4–5 uses or if it is greatly overheated or discoloured. Oils should be stored away from light to protect the vitamin E in them, which helps protect against possible harmful substances produced during cooking, especially frying.

Role of body fat

Body fat serves various functions. Some forms an integral part of the structure of cells, some acts as a protective covering for some organs (particularly the heart and the kidneys), some lies just beneath the skin and acts as an insulator and protects the underlying tissues from mechanical damage, some is stored in the fat depots as a source of energy and some lies between the cells of the tissues, especially muscle. In addition to providing energy, the essential fatty acids act as precursors of the very varied hormone-like substances known as prostaglandins.

Distribution of fat around the abdomen is a risk factor for cardiovascular disease.

Of all the body fat, the greatest variation in amount is seen with depot fat, which in adult men is normally about 8–15 kg and in adult women about 10–20 kg, although in the extremely obese it may reach as much as 100 kg. The distribution of depot fat is also very variable; in some people it is greatest in the limbs, while in others it is mostly around the abdomen. This latter distribution seems to be a risk factor for coronary heart disease. The depot fat and the subcutaneous fat are the first to be metabolized during weight loss, the fat covering the organs is used only during the final stages of starvation, while the fat that forms part of the structure of the cells is liberated only when the cells die.

Chapter 21

Alcohol

Ethyl alcohol (ethanol), abbreviated to alcohol in everyday language, has been known from time immemorial and has been used for the effect it has on the psychological state. Even small amounts depress the sense of anxiety and induce a sense of well-being and sociability. It is also valuable as an aid to sleeping if only small amounts are taken; larger amounts lead to disturbed sleep. In debilitated persons it often enhances the appetite as well as being a useful source of energy. Among its drawbacks are its induction of poor judgement and poor performance of skills requiring fine muscular movements and a loss of awareness of these changes. Of all the drugs of addiction in the western world, alcohol is by far the most abused.

Alcohol is rapidly absorbed into the blood from the stomach and the small intestine, requiring no digestion. It is metabolized mainly (over 90 per cent) by the liver and the products of this metabolism can then be used by the body generally. It yields about 7 kcal/g, thus giving more energy per gram than does carbohydrate or protein (4 kcal/g) but less than fat (9 kcal/g). As an energy supplier alcohol has a drawback because it causes considerable dilatation of the blood vessels of the skin, thereby allowing some of the energy it provides to be lost as heat. Because the extra blood flowing through the dilated vessels of the skin makes the subject feel warm, dangerous rapid cooling of the body outdoors may not be noticed.

Alcohol can easily cause overweight.

Alcohol cannot be stored by the body and is only trivially excreted in the breath and urine (about 5 per cent of the intake). It is metabolized at a rate of about 50–200 mg of alcohol/kg body weight/hour, the rate being more or less constant in any individual. An average 70 kg

It takes about 3 hours to clear all of the alcohol from the body after having a pint of beer.

person might metabolize about 7 g of (about 9 ml) alcohol in one hour. After having had one pint (600 ml) of beer, containing about 21 g (about 27 ml) of alcohol, it would therefore take about three hours to clear the body of all alcohol. Most people drinking four pints of beer or one bottle of table wine (700 ml) during an evening would still have alcohol in the blood next morning.

The sugar fructose has been used to treat patients in alcoholic coma because it increases the rate of metabolism of the alcohol but it can induce a dangerous rise in acidity of the blood (lactic acid acidosis).

Disulfiram (Antabuse: Dumex)

This substance is sometimes used to help alcoholics avoid alcohol. When alcohol is metabolized, the substance acetaldehyde is produced and it is normally broken down to harmless products as fast as it is formed. When disulfiram is given the enzyme which normally breaks down the acetaldehyde is inhibited and the accumulation of the intact acetaldehyde in the tissues rapidly causes severe nausea, headache, giddiness and sometimes vomiting. Most people who have taken an adequate dose of disulfiram and then also alcohol seldom repeat their mistake. A regular adequate dose of disulfiram thereby greatly assists the alcoholic in resisting drinking alcohol.

Effect of alcohol on nutrition

Alcohol can affect nutrition in several ways. Excessive spending on alcohol may leave insufficient money for an adequate diet, which is therefore likely to be poor in protein, vitamins, iron and calcium. The main vitamin deficiency is of folate, which sometimes leads to a macrocytic anaemia. There may also be a gastritis, especially with higher concentrations of alcohol, resulting in loss of appetite and perhaps poor digestion and absorption if the upper small intestine is also damaged. If there is gastric bleeding the body iron stores will be further diminished, though some chronic alcoholics retain excess iron, which damages the liver and the pancreas. The damage to the pancreas may lead to further excess iron absorption. Iron supplements should therefore not be given to chronic alcoholics without evaluation of their need for the metal. If a normal diet is eaten plus much alcohol, the extra energy intake will lead to obesity with its several disadvantages. Spirits (whisky, gin, brandy, rum) supply only energy but some beers may provide useful amounts of the vitamin B complex, though there are better and cheaper sources of these.

Nutritional deficiency in chronic alcoholics is seen only in those whose alcoholism is severe and who are in the lower economic groups. In the more

well-to-do segments of the population, alcoholism tends to produce not nutritional deficiency but overweight.

After a bout of hepatitis you should refrain from alcohol for at least six months.

It used to be thought that the damaged liver found in many alcoholics was caused by deficiency brought about by the alcoholism but that is often not so. The liver can be severely damaged by excess alcohol intake even when the general diet is good, so that eating well is no certain protection from the damage that excess alcohol can do to the liver and also to the pancreas. After a bout of hepatitis it is advisable to refrain from alcohol for at least six months after the liver has recovered its normal functions and even then the intake should be kept low.

The defects that arise from vitamin B group deficiency in alcoholism affect the peripheral nerves, muscles of the arms and legs, muscles that move the eyes, memory and other higher mental functions. The arms and legs are characteristically thin and weak. There is often also sensory impairment and numbness. In very severe cases complete paralysis of these muscles may occur in a few days, with recovery taking many months. The eyes may show involuntary rapid movements either from side to side or up and down and in some cases the eyes cannot be moved to focus on near objects. The mental disturbances may include agitation, hallucinations, poor memory for new information and sometimes old information, loss of time perception, apathy, loss of adequate speech and delirium tremens. Ability to walk is poor and balance can sometimes be kept only if the feet are widely spaced. The abnormalities are treated with high supplements of the vitamins of the B group; recovery is variable.

Reducing alcohol intake often lowers high blood pressure in mild sufferers.

Reducing alcohol intake often lowers both the systolic and diastolic pressures in people with mild high blood pressure (hypertension). This beneficial effect is usually seen within one month. It is not necessary for there to be a fall in body weight, but if such a fall occurs because of the reduction in alcohol intake then the lowering of the high blood pressure will be even greater, reducing still further the risk of coronary heart disease and stroke. This fall in blood pressure is reversed if the alcohol intake is increased to the original level.

There is evidence that a moderate intake of alcohol on a daily basis gives some protection against coronary heart disease.

There is evidence that a very moderate intake of alcohol on a daily basis (one or two glasses of wine per day) gives some protection against coronary heart disease. It is doubtful, however, whether people who do not drink regularly should be encouraged to do so because of the damage that they may do by becoming more than very moderate drinkers. It is often very difficult for some people to resist increasing their alcohol intake once they embark on a regular daily dosage.

There seems to be a causal relationship between alcohol drinking and cancer of the mouth, pharynx, larynx, oesophagus and liver. Smoking makes this worse. The effect seems to occur with beer and wine as well as with the distilled spirits.

The effects of alcohol during pregnancy and lactation are described in Chapter 3.

The energy supplied by some common alcoholic drinks is given in Table 10.

Table 10 Energy supplied by common alcoholic drinks

	Measure ml	*Alcohol content (%)*	*Energy supplied (alcohol plus carbohydrate) kcal*
Beer, various	300	4	100–120
Cider, various	300	4	100–120
Wine, table, dry	120	12	100
Wine, table, sweet	120	12	125
Sherry, dry	50	16	60
Sherry, sweet	50	16	70–80
Madeira	50	16	70–80
Vermouth, dry	50	15	55
Vermouth, sweet	50	15	65–75
Whisky, vodka, rum, gin, brandy	30	40	85

Chapter 22

Water

Under hot conditions or following exercise, survival without water can be limited to 24 hours.

The need for water takes precedence over the need for any other nutrient. Whereas a healthy adult can live for several weeks without food, survival without water is unlikely to exceed just a few days and under very hot conditions or if exercise must be performed this may be shortened to as little as 24 hours. There is only a small amount of reserve water in the body and the dehydration of the tissues, including the brain, that water deprivation causes soon produces collapse and death.

An adult in good health can lose about 3.5–7 litres of water before symptoms of dehydration set in. The subject weakens, loses concentration, becomes unco-operative and has sunken eyes and bluish lips. If no adequate amount of water is taken collapse soon occurs.

Excessive water loss in babies must be treated quickly.

Babies are especially susceptible to dehydration and must be kept well hydrated with about 150 ml of water/kg of body weight/day under conditions of normal water loss.

Daily water balance

The total body water in young men is about 60 per cent of body weight and for young women it is about 50 per cent (Table 11). It is greater in men because they have more muscle (rich in water) and less fat (poor in water) than do women. For both sexes the total body water falls with age, reflecting the fall in musculature and the increase in body fat. About one-quarter of body water is in the blood and in the fluids outside the cells, while about three-quarters is inside the cells.

Water intake

The water available to the body each day comes from that taken in liquids (about 1500 ml), the water in solid food (about 700 ml) and the water produced

Table 11 Body water

		Per cent (%) of body weight
Women	20–30 years	50
	60–80 years	45
Men	20–30 years	60
	60–80 years	54

There is more water in men because they have more skeletal muscle and less fat. The fall in body water with age follows the fall in skeletal muscle for both men and women.

Table 12 Water balance

Average daily water intake		
As liquid (5–6 cups)		1500 ml
In solid food		700 ml
From metabolism		300 ml
	Total	2500 ml
Average daily water loss		
In urine		1500 ml
From skin		500 ml
In breath		400 ml
In faeces		100 ml
	Total	2500 ml

In profuse sweating, water loss from the skin can be several litres.
These values are for a moderate climate and no disease.
There is a considerable individual variation.

when the food is metabolized within the cells (about 300 ml) (Table 12). This averages around 2.5 litres per day but is very variable. Thirst controls the intake when the amount taken is near the lower end of the range and the kidneys deal with the water taken in excess of need.

Water loss

Water is lost from the body each day in the urine (about 1500 ml), via the skin (about 500 ml), via the lungs (about 400 ml) and in the faeces (about 100 ml) (Table 12). This is on average about 2.5 litres per day and balances the water intake. In health, when intake rises, the loss increases to keep total body water constant, the kidneys acting as the controlling mechanism. A very pale urine is

Table 13 Water secreted during digestion

		Water secreted in millilitres (ml)
Saliva		500–1500
Gastric juice		1000–5000
Bile		100–1000
Pancreatic juice		700–1000
Intestinal juice		700–3000
	Total	3000–11500*

*Only 50–250 ml are lost in the faeces, all the rest being absorbed into the blood, except when there is diarrhoea.

about 99 per cent water plus 1 per cent salts; a dark urine is about 95 per cent water. A minimum of about 500 ml (nearly one pint) of urine is required each day in order to remove the normal waste products of metabolism. A low urine output is found when the water intake is low, during fevers, with marked sweating and for a few hours after strenuous exercise. In some kidney diseases the urine output may be much diminished, while in others it may be greatly increased. In diabetes insipidus, a relatively rare condition in which the pituitary gland secretes an inadequate amount of anti-diuretic hormone, the daily urine output can reach 20 litres per day.

There is a very large secretion of water into the gastro-intestinal tract each day (Table 13), virtually all of which is reabsorbed into the blood. Normally the faecal water is 50–250 ml per day, with the higher amounts being excreted when there is more dietary fibre. This loss in the faeces can go up to several litres a day in diarrhoea; the extreme example being cholera. Water lost in excess in this way, plus accompanying salts, must be rapidly replaced, especially in young children in whom diarrhoea and vomiting quickly result in dangerous dehydration; it is a common cause of death in many parts of the world.

Water replenishment for exercise

To combat the dehydration caused by sweating during prolonged vigorous exercise it is desirable to drink about 500 ml of water about 15 minutes prior to the start of the exercise. If the exertion lasts some time, it is advisable to drink about 250 ml every 15–20 minutes if the exercise allows this. Replenishing the lost water in this way diminishes the undesirable effects of dehydration and is better than trying to replace all the lost water in a short time after cessation of the exercise. If some litres of water are expected to be lost, it is prudent to get weighed before the exercise and then again after the exercise, the loss in weight showing how much water needs to be taken to bring the body fluid back to the normal resting state. Relying on thirst to do this may take several hours or even a day or two.

Sweat contains various salts, especially sodium chloride and it is advantageous to add some sodium chloride to the water of replenishment. Adding much sodium chloride, however, is likely to delay the emptying of the fluid from the stomach into the small intestine, thereby delaying absorption of the water into the blood. This delayed emptying will be more noticeable the more severe the exercise. Adding useful amounts of any sugar (glucose, fructose, sucrose) to the drinking water will also retard stomach emptying and delay rehydration of the body's tissues.

Ageing

With increasing age there is a gradual diminution in thirst so that less water is taken. There is also a gradual fall in the ability of the kidneys to concentrate urine so that urinary volume increases, enhancing water loss. Because of this, it is very important to make sure that older people drink enough to enable an adequate volume of urine to be produced.

Older people, especially if their mobility is reduced, try to drink less so as to be less inconvenienced by having to empty the bladder. It should be explained to them that this is not satisfactory.

Water intoxication

It is possible to drink sufficient water to cause drowsiness, coma and even death. It requires the intake of 10–15 litres taken over a few hours and it may occur without the sense of having drunk too much. Although the drinking of excess water is often the result of psychiatric disturbance or the taking of drugs, it can occur in normal people who do not realize that excessive intake can be dangerous. There is a report of a man who used cold water to relieve toothache and swallowed over 10 litres; he was unconscious for about two days but eventually made a complete recovery.

Chapter 23

Dietary fibre

Dietary fibre is the name given to a complex group of substances found only in plants. They are celluloses, hemicelluloses, pectic substances and lignin. Because the first three of these are carbohydrates, dietary fibre is sometimes called unavailable carbohydrate or non-starch polysaccharide. Dietary fibre from one species of plant may be unlike the fibre derived from a different species of plant and this variation may produce very different effects. Dietary fibre cannot be digested in the normal way by humans, but some is broken down and used by the micro-organisms in the colon. These micro-organisms plus any intact fibre form the bulk of the faeces.

Estimates of the amounts of dietary fibre in foods depend on the methods used for fibre analysis and hence the fibre contents of foods given in different reference tables may be very dissimilar. For example, older reference tables usually give the fibre content of wholemeal bread as about 3 g/100 g, whereas the current value is given as about 8–9 g/100 g. Most values for fibre should be considered guides rather than accurate numbers.

Dietary fibre intake

Dietary fibre intake in the United Kingdom in the 1980s probably averaged about 20 g per day, with most people being in the range of 10–30 g. About three-fifths of this came from vegetables, about one-third from cereals and about one-tenth from fruit. Vegetarians have a much higher fibre intake than the general population, consuming 30–40 g per day. They tend to get proportionately more of their fibre from fruit than do the non-vegetarians. In some parts of the world intakes are over 100 g per day. Dietary fibre, however, is not essential as Inuit living on a traditional diet eat no cereals, almost no vegetables and very little fruit.

Effects of dietary fibre

Bulking effect

Dietary fibre increases the bulk of the faeces and softens them because it takes up water. Such faeces are easily passed with minimum effort. These effects vary considerably with the source of the fibre and whether or not it was cooked. Cereal fibre is particularly effective in increasing faecal volume. Drying fibre usually reduces its ability to take up water, thereby lessening the valuable stool-softening and bulking actions.

A reduction in bulk of the intestinal contents usually results in the material passing more slowly down the tract (increased transit time) and there are fewer bowel evacuations. There is no clear-cut definition of constipation, of which there are many causes, but a reasonable working definition is that anybody who needs to strain to pass each stool is constipated. In otherwise normal people, dietary fibre is usually very effective in relieving constipation.

With some forms of fibre the increased faecal bulk is due to fibre, while with other types of fibre the increased faecal bulk is due to increased micro-organisms which have multiplied by metabolizing the fibre, perhaps almost completely. In both cases there is increased faecal water, either in the fibre or in the micro-organisms.

People consuming relatively large amounts of fibre (30–40 g/day) can expect to pass two or more soft motions per day, in contrast to people on a very low fibre intake (5 g/day) who are likely to pass one firm stool every two or more days, perhaps requiring straining. The lack of straining with a high fibre diet is of especial value for those with haemorrhoids. An adequate water intake is necessary for the fibre to operate satisfactorily.

Because of the very varied causes of constipation, not all sufferers are helped by extra fibre and in some cases it can be harmful.

Colonic gas (flatus)

The metabolism of dietary fibre by the micro-organisms in the colon produces gases (carbon dioxide, hydrogen, methane) which are partially absorbed by the blood and then given off at the lungs and breathed out, while the rest is passed via the rectum. When more fibre is eaten, the gas production increases, sometimes considerably. There may be mild abdominal discomfort, especially when first starting on a diet high in fibre. For these reasons there may be a reluctance to continue such a diet.

Effect on blood cholesterol

The soluble gel-forming parts of dietary fibre, particularly the pectins of fruits and the gums of vegetables and pulses, reduce bile acid reabsorption in the intestine so that it is excreted in the stools. This may lower the blood cholesterol level but the effect is variable, being most noticeable in people with an initially high cholesterol concentration. The less soluble forms of fibre, such as bran and cellulose, do not have this action on blood cholesterol levels.

Diverticular disease of the colon

This disease of the colon is common in countries where the fibre intake is low and it is rare in places with very high fibre intake. Increasing the dietary fibre for people with diverticular disease usually alleviates the condition. It seems possible that a high-fibre intake from an early age might prevent the disease from developing.

Irritable bowel syndrome

This common condition is often, but not always, helped by a natural high-fibre diet. Bran itself is not usually helpful and may make up to half the subjects worse.

Loss of minerals in the faeces

Several minerals, particularly calcium, zinc and iron, may be taken up by the dietary fibre in the intestine and excreted in the stools. This loss of minerals is unlikely to be important except when the intake of them is low or when the amount of fibre eaten is very large (over about 100 g per day). For most people adding very liberal amounts of bran to the diet is harmless, although adolescent girls and the elderly might be wise to use only moderate amounts, perhaps an extra 20 g of fibre per day added to an ordinary diet.

Weight control

A high-fibre diet may be useful for people attempting to lose weight. Food with a high fibre content takes longer to eat, has fewer calories per portion, may delay the onset of hunger and may reduce the efficiency of absorption of the meal as a whole. Its effect on weight loss, although likely to be only small, may nevertheless be worthwhile.

Effect on bowel cancer

A diet high in fibre has been thought to be associated with a lower incidence of cancer of the colon and rectum, possibly because the intestinal contents remain in the large intestine for a shorter time when much fibre is eaten, or because cancer-producing substances are taken-up by the fibre, thereby protecting the bowel wall. However, a 1999 report of a sixteen-year study in nearly 90,000 US female nurses aged 34–59 years did not find that a high-fibre diet reduced the incidence of colon cancer.

Designing a high-fibre diet

It is better to eat a diet that is naturally high in fibre than to eat a diet of refined food to which bran has been added. Adding bran does not provide all the substances that have been removed by the milling or refining procedures. The adding of cereal bran to the diet may be necessary when the volume of the diet has to be kept low or when a low-calorie diet is being taken.

A high-fibre diet should contain fibre from cereals, beans, lentils, vegetables and fruit. Obtaining all or most of the day's fibre from only one source is less satisfactory.

The fibre content of common foods is given in Table 14.

Table 14 Fibre content of common foods

High (More than 8 g/100 g edible portion)	Almonds, Brazil nuts, bread (wholemeal), dates (dried), figs (dried), hazelnuts, lentils, peanuts, raisins
Medium (3–8 g/100 g edible portion)	Bananas, beans, broccoli, Brussels sprouts, cabbage, carrots, maize, spinach, watercress
Low (1–3 g/100 g edible portion)	Apples, apricots (fresh), avocados, bread (white), cauliflower, celery, cornflakes, grapefruit, lettuce, mushrooms, oranges, peas, pears, plums, potatoes, prunes, rice (white), sweet peppers, tomatoes
Virtually none	Beef, butter, cheese, chicken (no skin), chocolate, cod, duck (no skin), egg, fats, haddock, ham, heart, herring, honey, kidney, lamb, liver, mackerel, milk, oils, pork, prawns, rabbit, salmon, tongue, turkey (no skin), veal

Chapter 24

Beverages

The need for water can be satisfied by drinking plain water but only a few people find this sufficiently interesting, the majority prefering to take flavoured drinks. These may be natural fluids such as the juices of crushed fruits; they may be infusions made by steeping leaves or seeds in hot or cold water; they may be concoctions manufactured by the soft-drinks industry; or they may be the products of the brewers and vintners. Which beverages are used depends on social and cultural habits, often greatly influenced by advertising.

Tea

Using the leaves of tea bushes to produce a beverage goes back thousands of years. In the early years there were many ways of making an extract but about 1550 AD the current method of making an infusion from whole or broken, withered and fermented leaves became the standard technique. It was at that time that western travellers to China discovered the habit of tea drinking. The Chinese were secretive about the art of preparing tea leaves and making an infusion but samples of tea were eventually brought to the western world and interest in tea drinking spread rapidly. For a while tea was very expensive and was used sparingly but it became much cheaper with the planting of tea gardens in India, Assam, Sri Lanka (Ceylon) and Indonesia. In Britain, tea for breakfast replaced ale in the eighteenth century. Today there are tea gardens in many parts of the world, growing a wide variety of tea bushes. As there are several different ways of preparing the leaves, the tea drinker has an ample choice of refreshing infusions.

Tea is taken partly because of its taste and partly because it is found to be refreshing. Relief of fatigue is often noted. This latter effect is brought about by caffeine, which is a mild central nervous stimulant. People vary a lot in their sensitivity to caffeine, the amount in one cup of average strength tea (about 60 mg) being enough to prevent sleep in some people, while having little or no effect on others. Under ordinary conditions, up to five cups of tea of average strength can be taken before the effect of the caffeine may become excessive. Too much caffeine may cause sleeplessness, restlessness, a sense of anxiety, an irregular pulse and a muscular tremor. It is rare, however, for caffeine intoxication to do lasting harm.

Tea and coffee drinking are rarely harmful.

Another important ingredient in tea is tannin, some teas containing much more than others. It is tannin which gives tea its astringency, produced by the tannins combining with proteins in the mouth. In some people, tea may cause abdominal discomfort, which may be due partly to the action of tannins on the surface protein of the stomach and partly to the stimulation of excess secretion of hydrochloric acid. When milk is added to tea the tannins combine with protein in the milk and thereby reduce the tea's astringency. The milk has no effect on the action of the caffeine.

In addition to its action on the central nervous system, caffeine has a weak action on the kidneys causing more urine to be produced (diuresis). This is in addition to the similar effect of the water of the tea so that some people find it necessary to pass urine quite soon after only one cup of tea.

It is rare for the average amount of tea to have a deleterious action on the heart.

The pleasant effect of tea quite often induces a habit of tea drinking but this should not be confused with an addiction because although deprivation of tea in a habitual tea-drinker may cause passing annoyance there are no harmful withdrawal symptoms as occur with true drugs of addiction.

Tea has no appreciable nutritional value apart from any milk or sugar which may be added to it.

Caffeine is secreted in the breast milk so nursing mothers should not drink more than 1–2 cups of tea or coffee a day.

Caffeine is secreted into breast milk, so that nursing mothers should not take more than 50–100 mg of caffeine per day, which is 1–2 cups of tea or coffee. If more tea or coffee is taken, the caffeine in the breast milk may be enough to cause a restless infant.

Coffee

As with tea, coffee has little nutritional value apart from any milk or sugar taken with it, but it is used because of a liking for its taste and its stimulant effect. Once again caffeine is the main cause of stimulation of the central nervous system, a cup of coffee providing 50–150 mg of caffeine depending on the type and amount of coffee used and the method of preparation of the infusion. Most people can drink about five cups of coffee per day before the effect of the caffeine may become excessive. Habituation occurs with coffee drinking but like tea habituation it is not a true drug addiction and there are no serious withdrawal symptoms.

In addition to caffeine, coffee contains tannins in amounts similar to those found in tea. Probably because of the tannins some people find that coffee, especially if taken without milk, causes abdominal discomfort. It is sometimes found that only tea or only coffee causes indigestion but not both, although the tannin consumption is similar.

Unlike tea, which does not stale easily, coffee loses its flavour rapidly when the ground beans are exposed to air.

Drinking coffee does not seem to have any link with coronary heart disease, angina or stroke. People who drink six cups of coffee per day (300–400 mg caffeine) have no greater risk than those who take only one cup. Nor does moderate coffee drinking adversely affect the blood pressure. In general, six cups of coffee per day should not be exceeded because the effect of the caffeine on the central nervous system may then become too great.

Fruit juices and soft drinks

> Fruit juices and soft drinks can damage teeth, especially in the young.

These drinks may be very acidic and contain much sugar. If taken several times a day, especially between meals, they may damage the teeth, particularly the first set in children. Drinking through a straw can reduce this damage and rinsing the mouth with water on finishing the drink is very useful. Fruit juices and soft drinks should never be added to a baby's bottle and pacifiers (dummies) never moistened with them.

Some fruit juices are valuable for their vitamin C and mineral content, though they lack most or all of the dietary fibre present in the original uncrushed fruit.

Soft drinks generally have little nutritional value apart from the energy they supply from sugar. This extra energy intake can be high and needs to be controlled in people prone to overweight. Fortifying a soft drink with glucose or other simple sugar (monosaccharide) rather than with ordinary table suger (sucrose) gives it no advantage in terms of energy yield.

The vitamin C content of fruit juice depends on the fruit originally used and the method of preparation of the juice and its storage. Orange and grapefruit juices are rich in vitamin C with about 40 mg per 100 ml; pineapple and tomato juices have about 20 mg per 100 ml; while apple juice has only a small amount of vitamin C. Some juices have vitamin C added during manufacture. The nutritional information box on the package should indicate clearly the actual vitamin C content of the juice.

Some soft drinks contain caffeine and its concentration varies between different brands. Generally, one ordinary can or bottle of a soft drink containing caffeine gives about the same amount of the drug as does a cup of tea or coffee. The actual amount of caffeine should be stated on the container.

A can or bottle of some soft drinks provides a similar amount of caffeine as does a cup of tea or coffee.

Decaffeinated tea and coffee

Most people drink tea and coffee not only because they are thirsty but because they enjoy the effect of the caffeine. If, however, there is undue sensitivity to caffeine, there are many brands of decaffeinated tea and coffee, which may be consumed as refreshing drinks without the stimulus that caffeine provides. There are several methods of removing caffeine from tea and coffee and there is little to choose between them; they all leave a very small amount of caffeine (0.1–0.2 per cent) but not enough to cause a noticeable effect. The process of decaffeination causes little or no change in the taste of the tea or coffee.

Alcoholic beverages

These are dealt with in Chapter 21.

Chapter 25

Cholesterol

Cholesterol is sometimes called a fat but it is not a fat; it is one of a group of complex molecules called steroids or sterols. In the pure state it is a colourless crystalline solid. For everyday nutritional purposes it is not necessary to know its chemistry.

> Blood cholesterol level is raised by saturated fat, especially from milk, cream, butter and cheese. There is a large genetic basis for the blood level.

Cholesterol is present in every cell of the body as well as in the blood. From it are produced the bile acids, the sex hormones, hormones of the outer layer (the cortex) of the adrenal glands and vitamin D. Only about one-third of the blood cholesterol is of dietary origin even on a high-cholesterol diet; the other two-thirds of the cholesterol is made by the body. Cholesterol can be made by all the tissues, though the liver has overall control. Its synthesis is increased when the diet is rich in saturated fat, causing the total cholesterol concentration in the blood to rise.

The cholesterol in the blood is joined to protein to form three main complexes described according to their densities and the main ones are known as high-density lipoproteins (HDL), low-density lipoproteins (LDL) and very low-density lipoproteins (VLDL). A high level of HDL-cholesterol in the blood is beneficial, whereas a high level of LDL-cholesterol is undesirable. Although the total cholesterol concentration in the blood is altered a little by increasing or decreasing the cholesterol in the diet, this is much less important than the effect of the dietary saturated fat.

Saturated fats vary considerably in their ability to raise the blood total cholesterol concentration. The saturated fatty acids of intermediate length (lauric acid, 12 carbons; myristic acid, 14 carbons; palmitic acid, 16 carbons) seem to be the only ones to raise appreciably the cholesterol level. This is of interest because the fatty acids in butter and butter-fat products are mainly of intermediate length and hence are likely to raise the blood cholesterol, unlike the longer fatty acids found in red meat. The fatty acids with chains of less than twelve carbon atoms have no effect on blood cholesterol concentration.

The mono-unsaturated oleic acid, abundant in olive oil, is as good as the polyunsaturated linoleic acid in lowering the undesirable LDL-cholesterol concentration. In addition, oleic acid does not reduce the desirable HDL-cholesterol concentration, which linoleic acid is likely to do.

Blood cholesterol and atheroma

> Plaques rich in cholesterol form on the inside of arteries. This starts in adolescence with an ordinary diet.

Atheroma is the condition in which plaques (lumps) of waxy material are found on the inside (lumen) of the arteries. These plaques are rich in cholesterol and tend to be larger and more numerous when the blood LDL-cholesterol level is high. The condition starts in adolescence on an ordinary UK diet and may be extensive by 30–40 years of age. These plaques, together with the blood clots which adhere to them, interfere, sometimes severely, with the free flow of blood through the tissues supplied by the damaged arteries. This occurs in arteries throughout the body but is of special importance in the brain and the heart, where an inadequate blood flow produces grave consequences. The laying-down of atheromatous material on the inner wall of the arteries is increased by a rise in the undesirable LDL-cholesterol level, but is decreased by a rise in the desirable HDL-cholesterol level. The HDL seems to be able to carry away cholesterol from the arterial wall and transport it to the liver where it is stored or metabolized. A rise in total blood cholesterol, therefore, is likely to indicate increasing risk of coronary artery disease only if the extra cholesterol is of the LDL type; if the rise reflects an increase in the HDL-cholesterol then the risk of arterial disease is actually diminished. *In summary*: a low total cholesterol concentration in the blood is good; a high total cholesterol concentration is bad only if the increase is because of a rise in the LDL-cholesterol; if the rise is due to an increase only in the HDL-cholesterol the situation is entirely satisfactory.

> A high total cholesterol concentration is bad only if the majority is LDL-cholesterol.

Reducing blood cholesterol

There are many causes of a raised total cholesterol in the blood and only a rise due to the diet is likely to respond substantially to a change in diet.

Increasing the concentration of the desirable HDL-cholesterol in the blood can often be achieved by taking regular moderate activity, such as a daily walk of 2–3 miles in about 45 minutes. If walking is not practical other forms of regular moderate activity can be substituted. Strenuous activity is not necessary and in

An increase in the HDL-cholesterol concentration of the blood can be achieved by moderate daily exercise.

some cases could be harmful. Other ways of reducing total cholesterol concentration in the blood and of sometimes raising the HDL-cholesterol are to:

a) eat fish and poultry (without skin) rather than red meat
b) trim off all visible fat
c) avoid fried food; eat boiled, baked or steamed food
d) limit the intake of cakes, biscuits, chocolate
e) use oils for cooking rather than butter, margarine or lard
f) avoid sausages, hamburgers, offal, minces, meat pies, pâtés, shellfish
g) avoid ice-cream made with milk fat
h) increase dietary fibre, especially from fruit
i) avoid food containing butter, palm oil, coconut oil
j) keep a slim figure (body mass index 20–25)
k) reduce smoking.

All plant foods are of value in reducing the blood cholesterol level because they contain a group of substances called sitosterols which reduce cholesterol absorption from the intestine into the blood. Sitosterol has been added to margarine and its use has been claimed to lower blood total cholesterol by about 10 per cent, enough to reduce coronary heart disease by about 30 per cent.

These recommendations should be adopted with moderation; do not go to extremes. For most people change in diet can achieve only modest effects on blood total cholesterol so that a burdensome alteration in diet may not be worthwhile. Keeping the total blood cholesterol level low should be the goal from an early age because lowering it after it has been raised for several years does not seem as satisfactory. The current view (1998) that the level in people over 30 years of age should not be more than 5.2 mM (200 mg cholesterol in 100 ml blood) puts about one-quarter of the UK's population in need of diminishing their cholesterol level.

There has been a suggestion that reducing the blood total cholesterol level may increase the risk of death from accidents and suicide but a large survey in 1994 did not confirm this.

A belief that eating garlic reduces the blood cholesterol has not been borne out by a twelve-week trial reported in 1998. Tablets containing garlic powder had no effect on the cholesterol level or on the triglycerides or lipoproteins.

Genetic influence

Genetic make-up has a large effect on the blood cholesterol levels. Two healthy individuals eating the same food and with very similar lifestyles may have quite different cholesterol levels. Also, changing the diet and life-style may have a beneficial action in one person but very little effect in another. Nevertheless, it is well worthwhile correcting the diet when there is a substantially raised LDL-cholesterol concentration.

Dietary sources of cholesterol

Nutritionally significant amounts of cholesterol are found only in foods of animal origin. The amounts of cholesterol in plant foods are extremely small.

The cholesterol content of common foods are given in Table 15. The daily intake in the United Kingdom averages about 0.4–0.7 g. The liver produces about 1.0–1.5 g of cholesterol per day, this being somewhat reduced by a high intake of cholesterol and is greatly increased by a high intake of saturated fat of medium chain length (lauric acid, myristic acid, palmitic acid).

Table 15 Cholesterol content of common foods

Very high (More than 250 mg/100 g edible portion)	Brain, butter, egg, kidney, liver
High (100–250 mg/100 g edible portion)	Cheese, crab, cream, heart, lobster, oyster, prawns, tripe
Medium (25–100 mg/100 g edible portion)	Beef, chicken (no skin), cod, duck (no skin), lamb, mackerel, mutton, pork, rabbit, salmon, veal
Low (Less than 25 mg/100 g edible portion)	Beans, cereals, egg white, fruit, nuts, peas, skimmed milk, turkey (no skin), vegetables, whole milk

Chapter 26

Vitamins: general

Vitamins are substances which are essential for health but which the body cannot make for itself either in sufficient quantity or at all. Vitamin D is an exceptional case as will be described later. Some vitamins are needed in only minute amounts, such as vitamin B_{12} (about 0.1–1.0 μg per day), while others are required in relatively large amounts, such as vitamin C (about 30–50 mg per day). Not all animals require the same vitamins.

During the early work on these substances, towards the end of the nineteenth century, their nature was unknown. What was clearly established, though, was that highly purified fat, carbohydrate, protein, minerals and water could not keep animals in good health unless some unrefined (natural) food was added to the diet. For example, as little as one teaspoonful of milk per day was enough to enable young animals to thrive on the experimental refined diet. The milk obviously contained one or more vital ingredients. It had been known for centuries that an unvaried diet containing only a few different food items was likely to be unsatisfactory, the best-known circumstance being the occurrence of scurvy, often fatal, among sailors on long journeys. This calamitous disease could be prevented or cured by quite small amounts of citrus fruit juice. In addition to scurvy, many other abnormal, sometimes fatal, conditions were produced experimentally in animals by the use of restricted diets and for each abnormality a cure could be obtained by adding small quantities of one unrefined food. Gradually these essential substances were isolated and their chemistry and functions described.

Nomenclature

The early nutritional experiments showed that vitamins fell into two broad groups: those that were soluble in water and those that were soluble in fat. Because it took many years before these active substances were purified and their chemical composition clarified, they were given letters as a means of identification. The water-soluble ones were B and C, while the fat-soluble vitamins were A, D, E and K. Research showed the C activity was due to only a single chemical, ascorbic acid. In contrast, A activity could be exerted by retinol, from animals,

Table 16 Names of vitamins

Type of vitamin	*Other name*
Vitamin A	retinol
Vitamin B_1	thiamine
Vitamin B_2	riboflavin
Vitamin B_6	pyridoxal or pyridoxine
Vitamin B_{12}	cobalamin
Vitamin C	L-ascorbic acid
Vitamin D	cholecalciferol or calcitriol or vitamin D_3
Vitamin E	alpha-tocopherol
Folate	folic acid or folacin
Niacin	nicotinic acid or nicotinamide

There are no usual other names for vitamin K, biotin or pantothenic acid

> If enough sunlight reaches the skin then the body can make some or all of the vitamin D it needs.

and by carotenes, from plants. Similarly, E activity was brought about by a group of substances called tocopherols and K activity was also the effect of several materials which had similar biochemical properties. Vitamin D activity was found to be more complicated because if enough daylight reached the skin the body could make some or all of the vitamin D it needed but, in the absence of adequate daylight, it was essential to have D-active substances from food.

The activity of the B group of vitamins turned out to be very complex because it contained several unrelated substances each of which had its own unique biochemical action. Some were distinguished by a number, such as vitamins B_1, B_2, B_6 and B_{12}, as well as their chemical name, but several were generally known only by their chemical name. Table 16 lists the UK names of the known vitamins.

Apart from vitamin D, which can be made as a result of daylight on the skin, all the vitamins are derived from plants or micro-organisms. These micro-organisms are generally outside the body, but there are some in the large intestine (colon) which produce vitamin K which can be absorbed into the blood in sufficient amount to satisfy the human daily requirement. The vitamins needed in the human diet come, therefore, from plants with their seeds and nuts, from micro-organisms in the food and from vitamins stored in the bodies of animals used in the human diet.

Vitamins may be ingested in their fully formed active state but sometimes they are generated in the body from related molecules. For example, vitamin A can be formed from the yellow and orange pigments, known as carotenes, found in plants, and niacin can be made from the essential amino acid tryptophan, present in the dietary protein. These related molecules are called pro-vitamins or precursors. Some vitamins can now be made in the laboratory.

Table 17 Relationship between international unit (i.u.) and weight of vitamin

Vitamin A	1 i.u. is 0.344 μg vitamin A acetate or 0.300 μg vitamin A alcohol (retinol)
β-carotene	1 i.u. is 0.60 μg β-carotene
Vitamin B_1	1 i.u. is 3 μg thiamine hydrochloride
Vitamin C	1 i.u. is 0.05 mg L-ascorbic acid
Vitamin D	1 i.u. is 0.025 μg cholecalciferol (D_3)
Vitamin E	1 i.u. is 1.0 mg DL-alpha-tocopherol acetate

International unit not defined by weight for vitamins B_2, B_6, B_{12}, K, folic acid, niacin or pantothenic acid.

Measures of activity

In the early days of study of vitamin activity investigators had available foods containing unknown amounts of unknown substances which they used to cure or prevent deficiency diseases. They described the substances under trial in terms of units of activity. Later, when the active substances were isolated the units of activity per gram of substance could be calculated. Table 17 lists the amount of each vitamin needed to exert one international unit (i.u.) of activity. No unit of activity has been defined for vitamins B_2, B_6, B_{12}, K, folic acid, nicotinic acid or pantothenic acid.

Toxicity

Large amounts of vitamins A and D can be fatal.

Large doses taken over many weeks seem to be harmless for vitamins B_1, B_2, B_{12}, E, K and for biotin, folate, niacin and pantothenate. In contrast, vitamins A, B_6, C and D produce toxic symptoms if taken in excess. Toxic effects do not occur when very large amounts of carotene are eaten because conversion to vitamin A is limited to only the amount of the vitamin actually needed. Too much carotene can harmlessly colour the skin, which resolves when the excessive carotene intake is stopped. This colouring of the skin may resemble jaundice but can be distinguished from that condition because carotene does not colour the sclera (the white of the eye) whereas in jaundice the sclera become yellow.

Deficiency diseases

Sometimes vitamin lack may cause an easily recognized deficiency illness such as scurvy (lack of vitamin C), pellagra (lack of niacin) or night blindness (lack of vitamin A). Generally, however, many tissues are unable to function properly during vitamin deficiency and then there is a malaise difficult to diagnose without a detailed knowledge of the diet taken.

Vitamin deficiency in the United Kingdom is rare in the absence of disease but it may occur as a result of food fads or when the diet is unvaried over a long period and consists of only a narrow range of foods.

Supplementation

> If you eat a normal varied diet you are unlikely to need vitamin supplements.

Healthy non-pregnant non-lactating adults eating an ordinary varied diet almost never need vitamin supplements. Many people, however, take vitamin supplements and provided that they do not exceed the recommended daily vitamin requirements by more than two-fold there is no danger. Taking very large supplements may be harmful. If it is known that the diet may be low in a particular vitamin then the diet should be improved or a supplement of about one daily recommended dose may be taken.

Increased vitamin intake will not make normal healthy children more healthy or more intelligent, nor will it give them more energy. In healthy adults, emotional stress is not helped by extra vitamins, nor is ageing deferred or general diseases diminished. In menstruating women, premenstrual symptoms are not helped by extra vitamins and high doses of vitamin B_6, sometimes used, can be harmful.

Some people may require regular vitamin supplementation, either for a short period or permanently. For example, the newborn need vitamin K; pregnant and lactating women may need vitamin B_{12} and folate; housebound individuals will need vitamin D; alcoholics may need vitamin B_1; vegans and especially their infants will need vitamin B_{12}. Others who may have an inadequate vitamin intake are the very poor, especially if very elderly; food faddists; convalescents, especially after surgery; slimmers on a very low energy intake and perhaps eating an unvaried and restricted range of foods; those with certain chronic diseases. For all these people, supplements should rarely exceed the recommended daily intake.

Recommended daily intake

The widely used recommended daily amount (RDA) is the amount of a substance needed by almost all the healthy members of the population with whom the RDA is concerned. Quite different groups of people might have different requirements. The RDA almost always has a safety margin included so that deficiency is very unlikely to occur. The RDA sometimes varies for different countries. In some countries, including the United Kingdom, there is a new series of values known as dietary reference values (DRV), with the RDA becoming the reference nutrient intake (RNI).

Cooking hints

Avoid vegetables which are old. Frozen vegetables are often a better source of vitamins than are unfrozen ones. Cook from frozen.

Eat food as soon as possible after cooking. Food kept on a hot plate or in an oven after cooking may lose almost all its vitamin C and other vitamins may deteriorate.

Place vegetables into boiling water rather than into cold for cooking. Leaching out of vitamins will be reduced. Steaming is better than boiling.

Mashing potatoes causes much of the vitamin C to be destroyed by the oxygen in the air.

Do not use copper vessels because the copper increases the rate of vitamin C destruction.

Cook vegetables until they are tender. Under-cooked vegetables pass through the intestine poorly digested.

Keep milk in the dark. Do not leave milk in bottles on the doorstep on summer mornings; have the milkman place the bottles in a covered box.

Chapter 27

Vitamin A

In the early study of nutrition it was found that young animals grew well when a few millilitres of fresh milk were added to an otherwise deficient diet. A fat-soluble substance was extracted from the milk and called vitamin A. Later, when purified, it was seen to be a colourless compound, stable at ordinary cooking temperatures but easily destroyed by air, by ultra-violet light (sunlight) and by immersion in rancid cooking fat. Subsequent analysis showed that vitamin A was retinol and that there were several related compounds in the body.

There is in plants a group of red and yellow pigments known as carotenoids, some of which can be converted to vitamin A. In their unconverted state they have no vitamin activity. The most important of the carotenoids is the red pigment β-carotene (beta-carotene); it is extremely abundant in carrots and occurs also in many other plants. It consists of two molecules of vitamin A joined together and is easily turned to free vitamin A during its absorption into the blood by the small intestine. This intestinal conversion is controlled and toxic amounts of vitamin A are not produced, so that even very large intakes of carotene are harmless. In contrast, when excess pre-formed vitamin A is eaten enough may be absorbed into the blood to be life-threatening. When there is a marked rise in carotene in the blood the skin may become coloured and resemble jaundice but the sclera (whites of the eyes) remain unchanged, whereas in true jaundice they become yellow.

> The absorption of vitamin A and carotene is poor if the diet is low in fat.

The absorption of vitamin A and of carotene is poor if the diet is extremely low in fat, a modest amount of which allows adequate uptake of both these substances.

In general nutrition, the term vitamin A is used to indicate vitamin A plus carotenoids.

Functions

Vitamin A is essential for the normal growth and function of most tissues, especially the eyes, the skin, the linings of the lungs, intestine and urinary tract, the

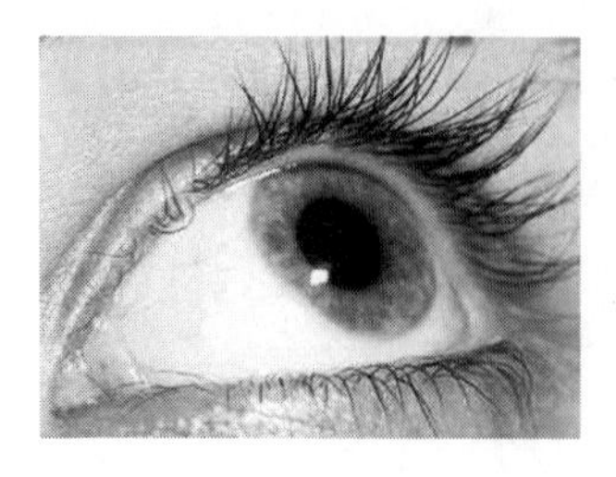

reproductive system and bone. In the young, growth ceases when the body is severely depleted in vitamin A.

One of the earliest observations was that vitamin A deficiency resulted in the inability to see adequately in poor light (night blindness). This is because the eyes become unable to produce sufficient light-sensitive rhodopsin (visual purple), the pigment in the cells (the rods) of the retina which are used at very low intensities of light. Daytime vision is normal because the retina's cone cells, which respond only to bright light, are then working. The satisfactory treatment of night blindness with liver has been known for thousands of years, though how it worked was, of course, not understood.

Vitamin A deficiency leads to night blindness.

Epithelial surfaces become deranged in the absence of enough vitamin A. The cornea of the eye becomes heaped-up and dry, interfering with vision. The problem is exacerbated by the failure of tear production, with consequent loss of lubrication and cleaning of the corneal surface. As this condition (xerophthalmia) gets worse the cornea becomes soft, infected and may ulcerate, with the whole eye in danger of destruction. Vitamin A deficiency is one of the commonest causes of blindness in parts of Africa where there is a cultural reluctance to feed green plant food to the very young, depriving them of carotenes, and as there is little food of animal origin pre-formed vitamin A is also lacking in the diet.

The skin shows marked changes in vitamin A deficiency, becoming dry and very rough, especially where there is pressure, such as over the shoulders and on the buttocks.

The lining of the respiratory tract becomes flattened, heaped-up and prone to infection. The action of vitamin A in preventing respiratory tract infection gave it its early name of anti-infective vitamin.

In the absence of adequate vitamin A during pregnancy the development of the embryo is impaired, leading to miscarriage.

Wound healing is always poor in people with vitamin A deficiency.

There have been claims that vitamin A reduces the incidence of cancer but a five-year trial of β-carotene, reported in 1990, found no such anti-cancer effect.

Sources

The vitamin A content of common foods is given in Table 18. Apart from liver, butter, cheese, eggs and kidney, animal foods have little vitamin A. When the fat is removed to make skimmed milk, almost all the vitamin A activity is

Table 18 Vitamin A plus carotene content of common foods

Very high (More than 2000 retinol-equivalents/100 g edible portion)	Carrots (2000), liver (20,000–30,000)
High (500–2000 retinol-equivalents/100 g edible portion)	Apricots (dried), butter (summer), dark leafy vegetables, margarine (fortified), spinach, watercress
Medium (100–500 retinol-equivalents/100 g edible portion)	Apricots (fresh), cheese, egg, honeydew melon, kidney, lettuce, peaches, prunes, tomatoes
Low (Less than 100 retinol-equivalents/100 g edible portion)	Bananas, beans, beef, cereals (unfortified), chicken, fish, ham, milk, oranges, peas, pork, turkey

removed along with the fat, so that the final product has to have vitamin A added.

In the United Kingdom, margarine has to be fortified to make its vitamin A content about equal to summer butter, which has more of the vitamin than has butter made from winter milk.

The livers of fish may be particularly rich in vitamin A and should not be eaten. Because fish meal is often fed to cattle and sheep, the livers of these farm animals may contain abnormally high vitamin A activity. As large doses of vitamin A may damage the developing baby, pregnant women and those who may become pregnant are advised not to eat liver or foods containing liver.

Fruits other than apricots contain little vitamin A activity and cereals are almost devoid of the vitamin. Dark-green leafy vegetables, on the other hand, are very good sources.

Cook plant food until it is tender.

Among everyday plant foods, carrots are by far the richest source of β-carotene. They should be cooked until they are soft. Five times more carotene can be absorbed into the blood from soft-boiled carrots than from raw sliced ones, which pass through the intestine more or less undigested. Blending carrots is especially good for carotene absorption.

Vitamin A and β-carotene are both stable to cooking at 100°C (boiling) but exposure to sunlight quickly reduces the vitamin activity. Milk in clear glass bottles should be kept shielded from light. Canning does not harm β-carotene and little is lost in the water despite its coloured appearance.

The average UK diet supplies about 1300 μg of retinol-equivalents per day and the liver of a healthy well-nourished adult usually has a 2–3 year store of vitamin A.

One microgram (1 μg) of vitamin A is equal to 3.3 international units.

Table 19 Vitamin A plus carotene satisfactory daily intakes

	Retinol-equivalents *µg*
6–12 months	300–1000
1–10 years	1000
Over 10 years	750
Pregnancy	750
Lactation	1200

One international unit is 0.3 µg retinol-equivalents

Recommended intakes

The amount of vitamin A needed each day is not known for certain because of the variable rates of absorption and conversion of the plant carotenoids and because of the uncertainty of which measure to use as a test. One measure is the prevention of night blindness. This requires about 450 µg of retinol-equivalents each day, which when doubled to give a margin of safety gives a value of about 900 µg per day or about 14 µg/kg body of weight. The average daily intake in the United Kingdom, about 1300 µg, amply satisfies this estimate. Another test is to measure the vitamin A activity in the blood of healthy subjects but the levels are very variable, with some apparently healthy people having quite low values. For infants, the recommended daily intake is based on the vitamin A content of breast milk. Table 19 gives intakes that are probably generally sufficient. Greater intakes consisting of carotene are harmless but large amounts of pre-formed vitamin A should be avoided. Vitamin A need in pregnancy and lactation is described in Chapter 3.

Toxicity

Excess vitamin A poisoning can occur and is usually the result of supplements such as cod liver oil.

Unlike most vitamins, excess preformed vitamin A is toxic, even fatal. The poisoning is almost always caused not by everyday food but by supplements such as cod-liver or halibut-liver oil taken when unnecessary and in excess of the recommended dose. Mistakes can be made by confusing milligram (mg) with microgram (µg): intakes of vitamin A are given in µg. Recovery is usually rapid (a few days) when the excess intake is stopped. The poisoned subject gets headache, loss of appetite, vomiting, diarrhoea and a rough, dry and sometimes peeling skin. The skin changes may look like those that occur in vitamin A deficiency and may therefore lead to a mis-diagnosis followed by a disastrous increase in vitamin A intake instead of the needed cessation. The cause of death is usually liver failure.

Chapter 28

Vitamin B_1

Vitamin B_1 is freely soluble in water so it can easily leach out during cooking.

Vitamin B_1 is a single colourless crystalline substance known as thiamine hydrochloride (it has also been called aneurine hydrochloride). It is present in all unprocessed plant and animal foods, especially in seeds and heart. It is stable in acidic water but is destroyed if the water becomes alkaline. Because it is freely soluble in water, prolonged boiling can leach out considerable amounts and one-quarter of it may be lost during ordinary cooking. Sunlight (ultra-violet light) decomposes it and tannins in tea, coffee and wine precipitate it so that it cannot be absorbed into the blood.

The body store of the vitamin is quite small, being only about 25–30 mg in a well-nourished adult, so that a regular intake is necessary. It takes part in the metabolism of carbohydrate by all the cells of the body in the reactions which yield energy.

Sources

All unprocessed foods contain vitamin B_1 but during processing some or all of it may be lost. For example, when cereals are milled for the production of white flour and white rice, the outer husks and the germ, which contain the vitamin, are removed. Modern methods of freezing, drying and canning food usually lead to only very little loss of the vitamin. Satisfactory sources are given in Table 20.

Enzymes in tea can break down vitamin B_1.

Tea contains a heat-stable enzyme which can break down vitamin B_1 and much of the vitamin in a meal can be destroyed if accompanied by several cups of strong tea. Tea taken apart from a main meal will do no harm.

One international unit is 3 μg of vitamin B_1.

Table 20 Vitamin B_1 content of common foods

High (More than 0.5 mg/100 g edible portion)	Brazil nuts, ham, harricot beans, heart, lentils, oats, pork, soybeans, walnuts, wheat (wholegrain)
Medium (0.25–0.5 mg/100 g edible portion)	Almonds, bacon, bread (white), egg yolk, hazelnuts, kidney, liver, peanuts, peas
Low (Less than 0.25 mg/100 g edible portion)	Asparagus, barley, beef, carrots, cauliflower, chicken, chicory, cod, duck, egg white, fruit, goose, lamb, lobster, maize, oyster, potatoes, rabbit, salmon, tongue, turkey, veal, watercress

Requirements

The amount of vitamin B_1 needed each day is related to the carbohydrate intake. For an adult consuming a mixed diet containing 2500 kcal about 1.5–2.0 mg of vitamin B_1 would be a generous supply. In general, about 0.4 mg of vitamin B_1 per 1000 kcal consumed is ample. Exceeding this intake is very unlikely to do any harm. The excess vitamin is rapidly excreted in the urine. Very rarely, large doses given by injection have caused severe allergic reactions.

Because the vitamin B_1 requirement is related to carbohydrate intake, much refined carbohydrate, poor in the vitamin, should not be given to a malnourished person unless the vitamin can also be given. Refined carbohydrate fed to somebody who has little vitamin B_1 in store may cause an acute deficiency state. If vitamin B_1 is lacking in the diet, fat would be safer than carbohydrate. Deficiency symptoms are unlikely to occur unless the vitamin B_1 intake is less than about 0.25 mg per 1000 kcal per day from carbohydrate.

Alcohol increases the need for vitamin B_1 and at the same time reduces its absorption from food by damaging the intestinal lining. In addition, many alcoholics have an inadequate diet with the result that vitamin B_1 deficiency in the United Kingdom is seen chiefly in alcoholics on low incomes.

Older people should increase their vitamin B_1 intake.

Older people need an increased vitamin B_1 intake, probably at least 2 mg per day, even if their energy intake is less than 2000 kcal per day. They become deficient more rapidly than do the young and respond more slowly when given vitamin B_1 supplements.

Table 21 gives the vitamin B_1 requirements at various ages.

Deficiency

Lack of vitamin B_1 in an adult causes loss of appetite, nausea, loss of weight and weakness, followed by polyneuritis with numbness and muscular paralysis. In

Table 21 Vitamin B_1 satisfactory daily intakes in milligrams

	mg
0–6 months	0.3
6–12 months	0.5
1–6 years	0.8
6–14 years	1.4
Adults	1.5–2.0
Pregnancy, lactation	2.0–2.5

the young there is almost complete stoppage of growth and death can be rapid. The condition is known as beri-beri. If oedema is also present it is called wet beri-beri. Mental disturbances are often present. Pure vitamin B_1 deficiency is uncommon because the other members of the vitamin B group, which mostly occur in the same foods, will also be lacking in the diet. The symptom complex may therefore be very complicated.

A common cause of vitamin B_1 deficiency is the consumption of white rice in communities where the diet is poor. Eating white-flour products does not cause vitamin B_1 deficiency in the United Kingdom because there is still some of the vitamin remaining in the flour and the rest of the diet will almost certainly provide all the vitamin needed.

Chapter 29

Vitamin B_2

Vitamin B_2 is insoluble in fats and only slightly soluble in water so that not much is lost during cooking.

Vitamin B_2, riboflavin, is a single substance which forms orange-yellow crystals. It has also been called vitamin G. It is very stable to heat, acids and air but is destroyed in a few hours by sunlight (ultra-violet light). It is insoluble in fats and only slightly soluble in water, so that not much is lost in cooking unless the food is made alkaline by the addition of sodium bicarbonate or similar material sometimes used to enhance the green colour of vegetables. Because of the deleterious action of ultra-violet light, milk, a useful source of the vitamin, should not be left in the sunlight in a clear glass bottle. When solutions of vitamin B_2 are viewed in daylight a yellow-green fluorescence is seen; this sometimes occurs with freshly passed urine which may contain appreciable amounts of vitamin B_2 after food fortified with the vitamin, such as breakfast cereals have been eaten. Some of the vitamin is stored in the liver, spleen and kidneys but most excess in the diet is quickly excreted in the urine.

Functions

Vitamin B_2 is present in all the cells of the body, where it forms part of several enzymes involved in the release of energy during the metabolism of glucose and fatty acids. It is also involved, with vitamin B_6, in the conversion of the amino acid tryptophan to the vitamin niacin. The enzymes which contain vitamin B_2 and those which contain niacin act together in energy release and hence a lack of either of these vitamins usually causes similar deficiency symptoms. Vitamin B_2 also acts in the conversion of folate to its active forms and as these are necessary for the synthesis of DNA, vitamin B_2 is involved in tissue growth and cell reproduction.

Sources

Vitamin B_2 is present in virtually all foods, with liver and kidney being especially rich sources (Table 22). In the United Kingdom it is only a very restricted food

Table 22 Vitamin B_2 content of common foods

High (More than 2 mg/100 g edible portion)	Kidney, liver
Medium (0.2–2 mg/100 g edible portion)	Almonds, cheese, chicory, duck, egg, goose, haricot beans, herring, lamb, lentils, mushrooms, oysters, pork, salmon, soybeans, spinach, veal, watercress
Low (Less than 0.2 mg/100 g edible portion)	Asparagus, beef, bread, Brussels sprouts, cauliflower, cereals (unfortified), chicken, cod, fruit, green beans, haddock, halibut, ham, lobster, milk, oats, parsnips, peanuts, peas, rabbit, turkey, walnuts

Vitamin B_2 toxicity does not occur because any extra is excreted in the urine.

faddist diet that is likely to be inadequate in the vitamin. In eggs it is divided about equally between the yolk and the white part: the colour of the yolk is due to carotenoids and not to vitamin B_2. Pasteurizing and drying milk have little effect on the vitamin B_2 content. Toxicity does not occur because any excess of the vitamin is rapidly and easily excreted by the kidneys.

Requirements

The requirement for vitamin B_2 is more closely related to body weight than to energy intake, although the need for the vitamin does increase with a rise in metabolism, whatever its cause. The values given in Table 23 are sufficient to saturate the body and hence produce excretion of the vitamin in the urine. For adults, intakes of about 0.5–1.0 mg per day produce less than 10 per cent in the urine, while with more than about 1.3 mg per day there is about 20 per cent or more of the intake in the urine. Hence something between about 1.0–1.3 mg per day would seem to be adequate but a margin of safety needs to be added. For older people, it is better to use the upper part of the range.

Table 23 Vitamin B_2 satisfactory daily intakes in milligrams

	mg
Infants	0.5
Children	0.5–1.5
Adults	2.0
Pregnancy	2.5
Lactation	3.0

These intakes will saturate the tissues

During periods of disease or after burns or surgery extra vitamin B_2 should be given.

Deficiency

Vitamin B_2 intake in children is often low.

Despite the important role of vitamin B_2 in cellular metabolism a deficiency of the vitamin does not cause a major disease and seems never to be fatal. After many months on a deficient diet there are only minor general symptoms such as muscular weakness, itching, dermatitis and other conditions which are often due to concomitant deficiencies of other vitamins of the B group.

Inadequate intake of vitamin B_2 appears to be relatively common, with 10–20 per cent of schoolchildren getting less than recommended amounts, however few seem to be inconvenienced.

Chapter 30

Vitamin B_6

Vitamin B_6 is a complex of several very closely related substances known as pyridoxine, pyridoxal, pyridoxamine and their phosphorylated derivatives. The biologically active form of the vitamin is pyridoxal phosphate, which acts as a co-enzyme for many intracellular metabolic reactions. For adequate formation of pyridoxal phosphate there has to be an adequate amount of zinc or magnesium; a deficiency of these two metals may therefore result in a pyridoxal-deficiency disorder.

Whereas the other members of the vitamin B complex are mainly concerned with carbohydrate and fat metabolism, vitamin B_6 is mainly concerned with protein metabolism, although it also plays a part in carbohydrate and fat usage.

Little vitamin B_6 is stored in the body; dietary excess is rapidly excreted in the urine in an inactive form.

The need for vitamin B_6 in pregnancy and lactation is decribed in Chapter 3.

Functions

The main functions of vitamin B_6 are in reactions which affect amino acids. When there is excess protein in the diet, unneeded amino acids can be turned into compounds which can enter the chain of reactions which liberate energy. These reactions require vitamin B_6, as do reactions which convert specific amino acids to essential substances. For example, tryptophan is converted to serotonin, tyrosine is converted to noradrenaline, and histidine is converted to histamine. Serotonin, nor-adrenaline and histamine are powerful regulators of many systems in the body. In addition, vitamin B_6 is required for the conversion of the amino acid tryptophan to niacin. Vitamin B_6 is also involved in the synthesis of RNA; in haemoglobin production; in antibody formation; in production of the elastic component in connective tissue, especially in the skin; in the production of insulin by the pancreas; in growth hormone production by the pituitary gland; and in the synthesis of cholesterol.

In carbohydrate metabolism, vitamin B_6 helps control the release of glycogen from the liver and the muscles; while in fat metabolism it is needed for the production of the essential polyunsaturated arachidonic acid from dietary linoleic acid and for the release of energy from fatty acids.

Table 24 Vitamin B_6 content of common foods

High (More than 0.5 mg/100 g edible portion)	Bananas, liver, rice (wholegrain), walnuts, wheat bran
Medium (0.25–0.5 mg/100 g edible portion)	Avocados, barley, beef, cabbage, chicken, cod, halibut, ham, kidney, lamb, maize, peanuts, pork, potatoes, rice (white), salmon, tuna, veal
Low (0.1–0.25 mg/100 g edible portion)	Bread (wholegrain), carrots, cauliflower, cheese, egg, heart, milk, peas, prunes, raisins, spinach, tomatoes
Very low (Less than 0.1 mg/100 g edible portion)	Bread (white), fruit

Sources

Most ordinary UK foods contain useful amounts of vitamin B_6; Table 24 gives common sources.

There may be considerable loss of the vitamin during cooking and food processing and even freezing foods, usually not harmful for vitamins, can result in a marked fall in the vitamin B_6 content. Milling cereals to produce white bread and white rice can remove almost all the vitamin.

In the United Kingdom the average mixed diet seems to supply just about enough vitamin B_6 and this should probably be increased to give a better margin of safety.

Requirements

The need for vitamin B_6 is related mainly to the protein intake, more being needed with higher protein consumption. For adults, about 0.02 mg of vitamin B_6 per gram dietary protein seems to be sufficient, so that for 100 g of protein per day, about an average protein intake, the daily vitamin B_6 requirement would be 2 mg. For older people, this should be increased to 2.5–3.0 mg per day. Table 25 gives the amounts desirable at various ages.

Table 25 Vitamin B_6 satisfactory daily intakes

	mg
0–6 months	0.3
6–12 months	0.5
1–12 years	0.5–1.2
12–50 years	1.2–2.0
Over 50 years	2.5–3.0

Contraceptive pills may increase the need for vitamin B_6.

Women using oestrogen-containing oral contraceptive pills need extra vitamin B_6, up to about 5 mg per day, because the oestrogen interferes with the metabolism of the amino acid tryptophan and its conversion to serotonin. Serotonin is an important transmitter in the brain and lack of it sometimes produces mood swings, especially depression.

Some other medicines, such as isonicotinic acid hydrazide (isoniazid), used in tuberculosis, and penicillamine, used in severe rheumatoid arthritis, diminish the activity of vitamin B_6 and for such patients 5–10 mg supplements of the vitamin are desirable to prevent damage to the peripheral nerves.

Deficiency

In the early 1950s some babies fed an infant formula providing less than 0.1 mg of vitamin B_6 per day developed convulsion-like seizures which were rapidly cured by injections of the vitamin. Apart from these infants, no clear-cut illness has been identified as being due solely to vitamin B_6 deficiency. In adults a form of anaemia, very similar to that produced by lack of iron, may occur; it is rapidly cured by a vitamin B_6 injection but not by extra dietary iron.

Among the general non-specific findings in adults living on a diet low in vitamin B_6 are insomnia, irritability, dermatitis around the eyes and mouth, muscular weakness and difficulty in walking. These symptoms can sometimes be relieved by pyridoxine when the other members of the B_6 complex are ineffective.

Different species of animals sometimes show quite different responses to lack of vitamin B_6 and the results of such experiments cannot always be applied to humans. This is a common problem with work on the vitamins.

In the United Kingdom, illness due to lack of dietary vitamin B_6 is extremely rare.

Toxicity

Vitamin B_6 supplements should not exceed 10 mg per day.

Even though vitamin B_6 is water-soluble and can be excreted easily in the urine, toxic effects are produced by the taking of large doses of the vitamin over several months. For example, doses over 50 mg per day can cause severe damage to the sensory nerves in the arms and legs. The damage usually clears up on cessation of the supplementation. In July 1998 the UK government's advice was that supplements of vitamin B_6 should not exceed 10 mg per day.

Chapter 31

Vitamin B_{12}

Vitamin B_{12} is the name given to a group of substances called cobalamins. They are the largest molecules that can be absorbed into the blood by the small intestine and only about one-millionth of a gram (1 μg) or less is needed in the diet each day to prevent the fatal disease called pernicious anaemia. If the vitamin is injected, instead of eaten, then only 0.1 μg is required.

Up to 5 years supply of vitamin B_{12} can be stored in the liver.

The intestinal absorption of ordinary dietary amounts of vitamin B_{12} takes place in the last 30 cm or so of the small intestine (the end of the ileum) and for this to occur the vitamin has to combine with a substance called intrinsic factor (another very large molecule) and calcium ions. The intrinsic factor is secreted into the gastric juice by the stomach and it is almost always the lack of intrinsic factor, rather than lack of dietary vitamin B_{12}, which produces pernicious anaemia. The absorbed vitamin is stored in the liver, a well-nourished adult having up to 5 years supply. In the absence of intrinsic factor, enough vitamin B_{12} can be absorbed if very large doses of the vitamin are fed, which explains why eating about 200 g (7 ounces) minced raw liver per day cures pernicious anaemia. This treatment was replaced by the injection of the vitamin following its purification. The vitamin in its pure state forms small dark red crystals containing about 4 per cent cobalt. It is now obtained from a mould (*Streptomyces griseus*) grown for streptomycin production.

The absorption of vitamin B_{12} by the intestine is normally very efficient (90 per cent) when a small amount (0.5 μg) is in the diet but this rapidly falls as the intake rises, so that only about 1 per cent is absorbed for intakes greater than 50–100 μg.

Functions

Vitamin B_{12} is needed for the normal growth of all cells. It is involved, with folate, in DNA synthesis and in the metabolism of carbohydrate, protein and fat. It is needed also, without folate, for the production of the myelin sheath around nerve fibres.

In the absence of enough vitamin B_{12} the bone marrow produces abnormally large red cells in the blood, called megaloblasts (macrocytes). As the condition progresses the blood becomes deficient in haemoglobin and the result is a megaloblastic anaemia. The other blood cells are also diminished in number.

In the nervous system, lack of vitamin B_{12} causes severe degeneration of the spinal cord, due partly to inadequate production of the myelin sheath around the long fibres and partly to interruption of carbohydrate metabolism on which the nervous system relies almost entirely. If large doses of folate are given without vitamin B_{12} in pernicious anaemia, the available vitamin B_{12} is diverted to improving blood cell formation and thereby worsens the degeneration in the nervous system. The damage done to the nervous system by vitamin B_{12} deficiency is permanent, whereas the blood changes are reversible when sufficient vitamin is supplied.

Sources

> There is virtually no vitamin B_{12} in plants so vegans do not get any in their normal diet.

All vitamin B_{12} comes from micro-organisms (bacteria; moulds). People get vitamin B_{12} by eating contaminated plants and by eating animals which have stored the vitamin. Although the micro-organisms in the human colon also produce the vitamin the colon cannot absorb it and it is lost in the faeces. Plants themselves do not utilize vitamin B_{12} and they do not store it, so that well-washed plant food is usually devoid of the vitamin. For this reason vegans get virtually no vitamin B_{12} in their ordinary diet (Chapter 10).

Almost all food of animal origin supplies enough vitamin B_{12} for ordinary requirements, the best sources being liver and kidney, and very good sources are meat, poultry, fish, milk, cheese and eggs.

The UK mixed diet varies widely, containing about 1–100 μg vitamin B_{12} per day. The vitamin is reasonably stable to ordinary cooking, although its resistance to heat is reduced considerably by the presence of vitamin C.

Requirements

As the daily requirement of vitamin B_{12} is extremely small, even the poorest mixed diet is likely to provide enough for health. Few diets in the United Kingdom have less than about 3 μg per day. Some people appear to maintain health on as little as 0.5 μg per day. Experiments with radioactive vitamin B_{12} show that most adults need about 0.5–1.0 μg per day in the diet. Because the efficiency of absorption from the intestine falls rapidly as the intake of the vitamin rises, there is little advantage in taking large doses.

Disease resulting from vitamin B_{12} deficiency is almost never due to dietary lack of the vitamin but is caused by some other problem.

The requirements for vitamin B_{12} in pregnancy and lactation are described in Chapter 3.

Chapter 32

Vitamin C

Although the disease scurvy has been recognized for centuries, it was only when long sea journeys were undertaken that it became of major importance. On these voyages the loss of one-third of the crew from scurvy was not uncommon because the main foods were bread, salted meat and old potatoes, all of them virtually devoid of vitamin C.

By the early seventeenth century it was known that fruit juice prevented scurvy and a modern-style controlled experiment was carried out by James Lancaster on a journey to India. Some of his sailors were fed bottled lemon juice and they escaped scurvy, while many of the sailors without the juice died of the disease. Despite this evidence, as long as one hundred years later about six hundred sailors out of about nine hundred died of scurvy on a round the globe voyage with the Royal Navy. It took a further fifty years before the Royal Navy introduced fruit juice, particularly limes and lemons, for all sailors and thereby more or less banished the disease among their crews. Nevertheless, scurvy was a common and serious illness in both armies in the American Civil War (1861–1865) and fifty or so years later still Captain Scott and his team suffered from scurvy on their disastrous journey to the South Pole in 1912.

Vitamin C was eventually isolated in 1928 and shown to be a single substance: it was given the name ascorbic acid because of its anti-scurvy action. Within the next four years ascorbic acid was proven to be the anti-scurvy agent in fresh fruit juice, especially from oranges and lemons. There are two forms of ascorbic acid: L-ascorbic acid has vitamin activity but D-ascorbic acid does not. Among mammals, only humans, monkeys and guinea-pigs need vitamin C in the diet; all other mammals tested can make it.

As well as being an essential item in the human diet, vitamin C is used in the food industry in the making of bread and dough products; in frozen desserts; in jams and jellies; in soft drinks; in wine and beer production; and in the pickling and curing of meat.

Functions

Vitamin C has many important functions but its most obvious and best known is the prevention of scurvy. This condition, now rare in the United Kingdom, is brought about by a diet containing very little or no fresh fruit or vegetables. It occurs mainly in alcoholics, in debilitating illness, in food faddists and in the very poor. There is bleeding into the skin, which at first forms very small red spots (petechiae) and later large dark bruises. The gums become spongy and bleed and the teeth become loosened and may fall out. New wounds hardly heal, if at all, and old scars may break down. Bleeding into joints causes them to swell and be very painful. There may be bleeding into any tissue, with accompanying symptoms and death may occur by sudden collapse. The development of scurvy is very slow and may take several months. The disease process is the result of failure of the tissues to produce enough of a protein called collagen, which forms the support of blood capillaries, hence there is bleeding and lack of healing. Scurvy is also associated with anaemia, partly due to bleeding, partly to a lack of red blood cell production and partly to the inability to absorb enough dietary iron into the blood.

Despite many investigations, there is no evidence that vitamin C is particularly helpful in the common cold. Nor is there evidence that vitamin C has any anti-cancer value in established cases. It may, however, aid the prevention of initiation of cancer by helping to destroy some cancer-producing substances. Vitamin C supplementation in elderly people does not appear to increase their life span.

Sources

Most mixed diets in the United Kingdom contain at least 25 mg of vitamin C per day, which is about twice the amount needed to prevent scurvy. Good sources (Table 26) are oranges, cabbage, Brussels sprouts, broccoli and cauliflower but variable amounts are lost in cooking. There is practically no vitamin C in unfortified cereals, milk, milk products, eggs, meat, poultry or fish.

The amount of vitamin C in any plant depends on the conditions during its growth and how it has been stored. For some plants, length of storage time is very important. For example, green beans often lose a lot of their vitamin C in just a few hours after harvesting, while potatoes may retain the vitamin for months. Frozen foods retain vitamin C for long periods. High temperatures and bright light hasten the loss of the vitamin.

> Cook vegetables and fruits in water that has boiled for one minute.

To preserve vitamin C when cooking, the food should be added to water which has boiled for a few minutes to drive off the dissolved oxygen. Rapid heating of the food quickly destroys the enzymes in the

Table 26 Vitamin C content of common foods

High (More than 50 mg/100 g edible portion)	Broccoli, Brussels sprouts, cabbage, cauliflower, oranges, spinach, strawberries, sweet peppers, watercress
Medium (25–50 mg/100 g edible portion)	Bananas, gooseberries, grapefruit, lemons, limes, liver, melons, peas, raspberries
Low (10–25 mg/100 g edible portion)	Apples, avocados, blackberries, maize, pineapple, potatoes, tomatoes
Very low (Less than 10 mg/100 g edible portion)	Beans, beef, butter, carrots, celery, cereals (unfortified), cheese, chicken, egg, fish, grapes, ham, lentils, lettuce, milk, nuts, pears, plums, pork, turkey

cells before they break down the vitamin. The food should be eaten soon after cooking or rapidly cooled and frozen. Copper cooking vessels should not be used and sodium bicarbonate should not be added.

Lemon, grapefruit and orange juices are very acidic and can damage the teeth if there is prolonged contact. They are best taken only once a day, preferably with a meal.

Requirements

The amount of vitamin C needed to prevent scurvy in a healthy adult is only 10–20 mg per day. During illness, after surgery and in other stressful conditions the need for the vitamin rises, perhaps to as much as 100 mg per day. For most healthy people the body seems to be saturated with the vitamin when about 75 mg per day are taken. It may be that the aim should be to keep the tissues saturated, rather than to merely ward off scurvy. Some authorities recommend relatively low intakes while others are more generous. Table 27 gives intakes which are likely to keep the tissues saturated in most people. Excess vitamin C is easily excreted in the urine.

Table 27 Vitamin C satisfactory daily intakes in milligrams

	mg
Infants	35
1–10 years	35–60
Over 10 years	75
Pregnancy	100
Lactation	150
During illness	100

In a well-nourished adult, the body pool of vitamin C, stored mainly in the liver, is about 1500 mg but this can be increased to about 5000 mg on a diet rich in the vitamin. The average body store will prevent scurvy for about two to six months on a diet very poor in vitamin C.

Megadoses of vitamin C

The taking of daily doses of 1000 milligrams (i.e. 1 gram) of vitamin C has been popularized to cure or prevent a variety of illnesses. Most (50–75 per cent) of such doses are quickly excreted in the urine, either as unchanged ascorbic acid or as its metabolite oxalate. The oxalate usually causes no trouble but in susceptible individuals may produce stones in the kidneys and urinary bladder, although this is rare. Sometimes these megadoses cause diarrhoea and they may interfere with the absorption into the blood of vitamin B_{12}. Low blood sugar and anaemia have also been reported. There is no convincing evidence that megadoses are in any way more useful than normal dietary intakes.

Chapter 33

Vitamin D

The most obvious sign of a chronic lack of vitamin D is mis-shapen bones. There is bowing of the legs, bossing (bumps) of the forehead, lumpy ends of the ribs where they meet the sternum (front of the chest) and bending inwards of the lower ribs. There may also be hidden deformation of the pelvic bones, which may make normal delivery of a baby very difficult or impossible. In addition, the teeth often erupt late, are poorly calcified and decay rapidly. General growth is poor and stature is diminished. One or all of these abnormalities may occur in the condition called rickets, which presents itself primarily in young children.

Rickets has been known for thousands of years, although its cause has been understood only since the first quarter of the twentieth century. It became increasingly prevalent in the United Kingdom during the mid-seventeenth century and was very bad for the next three hundred years. The precipitating cause of the epidemic was the growth of towns with pollution of the atmosphere by smoke and dust and the shift of increasing numbers of people from outdoor to indoor work, all of which deprived much of the population of the action of daylight on the skin. Children of the rich living in towns suffered as much as the poor; indeed, as the latter spent most of their time in the streets they sometimes suffered less badly. By the mid-nineteenth century it was known that cod-liver oil could cure or prevent rickets and in the early 1800s ultra-violet light was shown to cure the disease, although this treatment was generally ignored for the next hundred years or so. This was especially unfortunate because the general diet was inadequate in vitamin D-active material and sunlight or ultra-violet light was the best and natural treatment. Only since the isolation and purification of vitamin D-active substances in the early 1900s has fortified food been able to safely replace sunning of the skin.

Sunning of the skin

When sunlight or ultra-violet light hits the skin it converts a cholesterol metabolite called 7-dehydrocholesterol to a related compound called cholecalciferol, also known as vitamin D_3. This then passes in the blood to the liver where it is slightly modified again to form calcidiol and this then travels via the blood to the

Exposing the skin to daylight is the best way to get vitamin D. Direct sunlight is not necessary. For many people prolonged exposure to direct sunlight is harmful.

kidneys where a further modification changes calcidiol to calcitriol. Calcitriol then passes via the blood to the target tissues, where it exerts its effects. This is, of course, a classic description of a hormone, which is a substance produced exclusively by a group of special cells and which is then carried by the blood to target cells where it exerts its special action. The main target cells for calcitriol are those lining the small intestine, particularly the upper part, and the bone cells. Calcitriol is therefore really a hormone rather than a vitamin. Nevertheless, it is described as a vitamin partly for historical reasons and partly because in the absence of adequate sunning of the skin cholecalciferol and a similar substance called ergocalciferol (vitamin D_2) can be obtained from food. Ergocalciferol is formed by the action of ultra-violet light on yeasts and fungi and is accumulated by plants and animals. When these plants and animals are eaten by people, the ergocalciferol undergoes the same changes in the liver and the kidneys as does cholecalciferol and the final form acts as does calcitriol. For ordinary everyday nutrition all these substances and some similarly active sterols can be referred to simply as vitamin D.

The amount of skin sunning required to produce enough vitamin D depends on the intensity of the light, the length of sunning and the colour of the skin. On a bright summer's day in the United Kingdom, sunning the face (without make-up or screening lotion) and hands for about 2–3 hours produces enough vitamin D (probably about 10 μg) to prevent rickets. Most people get more than this skin exposure to the sun during the brighter weather and are able to store enough vitamin D, mainly in their depot fat but a little in the liver, to tide them over the winter months. People with dark skins need longer periods of sunning, especially for heavily pigmented skin. Dark skin is not, however, a serious problem except in those cultural groups whose members, especially the women, keep virtually all the skin covered when out of doors. Thus the incidence of rickets in even lightly coloured Asians in the United Kingdom is undesirably high, partly because of the lack of skin sunning and partly because their diet contains too much phytate (mainly in chapati flour), which reduces the intestinal absorption of calcium. The much more heavily pigmented groups of African origin in the United Kingdom rarely suffer from rickets because they expose their skin to daylight sufficiently and have less phytate in their food. White-skinned nuns who cover their skin completely and white people who rarely go out of the house also have very low levels of vitamin D in their tissues and may suffer from deficiency of the vitamin. In contrast, white people exposed to the tropical sun for prolonged periods may sometimes produce excess vitamin D, enough to induce symptoms of vitamin D poisoning.

Functions

Vitamin D's main function is to metabolize calcium.

The main functions of vitamin D are concerned with the metabolism of calcium. It enhances the absorption of calcium into the blood by the small intestine, especially at its upper end, and it reduces the amount of calcium that is lost in the urine. In this way it increases the calcium and also the phosphorus (as phosphate) in the blood and tissues and helps bring about the laying down of calcium phosphate salts in bone, during both early growth and the re-modeling of bone in the adult. Without adequate vitamin D the bones are unable to withstand the forces acting on them, resulting in the deformations seen in rickets in the young and in osteomalacia (softening of the bones) in the adult. For normal action, vitamin D needs adequate levels of parathormone produced by the parathyroid glands (in the neck) and of calcitonin from the thyroid gland. Both these substances are hormones and their production is to some degree affected by the activity of vitamin D.

By helping to keep the amount of calcium in the blood from falling to a low level, vitamin D prevents convulsions and involuntary contraction of muscle (tetany).

Sources

Apart from eggs and fatty fish such as herrings, mackerel, pilchards, tuna and sardines, few normal foods in the United Kingdom have a useful amount of vitamin D unless they have been fortified. Some values are given in Table 28. There is no vitamin D in unfortified cereals, fruit, and vegetables, while meat, poultry and white fish have only trivial amounts. The most convenient source of vitamin D in food is fortified breakfast cereal, though the amounts added vary widely. The cheapest and perhaps best way to obtain the vitamin is to sun the skin regularly.

Table 28 Vitamin D content of common foods

	μg/100 g edible portion
Herring	100–1200
Salmon	200
Tuna, sardines, pilchards	8
Liver	5
Egg, veal	4
Butter (summer)	1
Cheese (hard)	0.25
Milk (unfortified)	0.03

If concentrates of fish liver oils are taken as supplements it is essential not to exceed the recommended dose because these preparations sometimes contain a great deal of the vitamin. The fish get the vitamin from surface plankton.

The daily vitamin D intake in an average UK diet is about 2.5–3.0 μg per day.

Breast milk has only 0.01–0.25 μg vitamin D per 100 ml, so that a daily intake of 800 ml would provide the baby with only 0.08 μg in the winter months and 2 μg during the summer time. This is an inadequate intake, especially during the winter when sunning the skin might also be inadequate, so that vitamin D supplements are necessary. Infant formula milk and follow-on formula milk are fortified.

Recommended intakes

The actual requirements of vitamin D are not known because it is uncertain how much of the vitamin is obtained from sunning the skin. However, a dietary intake of about 10 μg per day will prevent rickets in children and osteomalacia in adults and is a safe amount. In some people intakes as low as 2–3 μg per day seem to keep them healthy. Many children in tropical areas get virtually no dietary vitamin D and remain free of rickets because they get ample sunlight on their skin.

It is undesirable to exceed a dietary intake of 25 μg per day.

One μg is 40 international units.

Toxicity

Vitamin D is toxic in large doses. It is unwise to exceed about 25 μg per day; more than 45 μg per day have been found to retard growth in bone. Despite this, some children in the United States have a daily intake of about 60 μg and seem to be healthy; there must be a wide variation in sensitivity to the vitamin.

Marked excess of vitamin D, more than about 50 μg per day, may cause an excessive rise of calcium in the blood (hypercalcaemia), resulting in loss of appetite, headache, irritability, depression, confusion, coma and even death. If the hypercalcaemia is less marked but of many months duration there may be the laying down of calcium salts (calcification) in the heart, kidneys and large arteries, leading to their failure. This excessive intake is rarely from normal food but usually comes from larger than recommended doses of vitamin D supplements such as cod-liver oil or halibut-liver oil. These should not be taken without expert advice.

Chapter 34

Vitamin E

Vitamin E consists of a group of closely related yellow oils, the most active of which in ordinary food is alpha-tocopherol (α-tocopherol). They occur in many vegetable oils, especially seed oils. Common sources are given in Table 29. As well as being abundant in seeds, some vitamin E is present in every cell in both plants and animals.

Functions

During early work on vitamin E, the most impressive observations were miscarriages in female rats and sterility in male rats, which suggested that the vitamin's main function was in reproduction. It is now known, however, that it is vital for all cells, where it protects the cell walls (plants) or outer membranes (animals) from damage by oxidation. It also prevents oxidative change in polyunsaturated fatty acids by substances called free radicals; the need for vitamin E increases with a rise in the amount of polyunsaturated fatty acids in the diet. It also protects vitamin A. Vitamin E is itself oxidized during its protective action but it can be

Table 29 Vitamin E content of common foods

Very high (More than 20 mg/100 g edible portion)	Cottonseed oil, maize oil, peanut oil, safflower oil, sunflower oil, walnuts
High (10–20 mg/100 g edible portion)	Cashews, peanuts, soybean oil
Medium (5–10 mg/100 g edible portion)	Almonds, chocolate, coconut oil, olive oil, spinach
Low (1–5 mg/100 g edible portion)	Broccoli, butter, cheese, egg, kale, liver, oats, peas, rice (wholegrain), sweet peppers, veal, wheat (wholegrain)
Very low (Less than 1 mg/100 g edible portion)	Apples, bananas, cabbage, carrots, cauliflower, celery, chicken, grapefruit, haddock, ham, kidney, lettuce, maize, milk, onions, oranges, pork, potatoes, rice (white), tomatoes, wheatflour (white)

returned to its original form if there are sufficient vitamin C and other anti-oxidants present.

There is no good evidence that large doses of vitamin E delay ageing in people, although it has been used in that hope.

Human deficiency of vitamin E occurs virtually only when there is inadequate absorption of fat, or in premature babies, who absorb fat very poorly. The tissues to suffer most are the red blood cells, which breakdown easily, and the lungs, which are especially exposed to oxygen and other oxidizing gases such as ozone.

Sources

Merely measuring the total tocopherol content of a food does not give sufficient information about the vitamin E activity of the food because alpha-tocopherol is especially active while some other tocopherols have only little activity. For example, in maize oil only about 10 per cent of the total tocopherol is alpha-tocopherol, but in cottonseed oil it is about 60 per cent and in safflower oil about 90 per cent. Table 29 gives the alpha-tocopherol content of foods and not the total of the various forms.

Vitamin E is very stable, even at temperatures above 100°C, provided there is little oxygen. Hence slicing, shredding and blending, which allow air to reach many cells, encourage destruction of the vitamin, whereas roasting, baking and boiling uncut food do much less harm.

During commercial purification of plant oils there can be substantial loss of vitamin E activity and the same problem arises with the commercial preparation of food. Assessment of vitamin E activity should be carried out on the final product rather than on the food constituents before processing. Long storage of food, even in a refrigerator, can deplete it of vitamin E.

Requirements

Because vitamin E deficiency in humans, even on a meagre diet, is virtually unknown, daily need for the vitamin has been assessed on relatively few cases. Table 30 gives satisfactory intakes for various groups. In the United Kingdom

Table 30 Vitamin E satisfactory daily intakes

	mg
Infants	2*
Children	3–7
Adults	10
Pregnancy	10
Lactation	11

*Supplied by breast milk during first six months of life

there is a very wide variation in vitamin E in the diet, with values lying between about 10–60 mg per day. In normal healthy people there is never a need for vitamin E supplementation.

Vitamin E is remarkably safe: people have taken 1000 mg per day or more for many months without apparent harm. Excess vitamin E is stored in the liver and in the depot fat. This is usually sufficient for several months on a diet very low in the vitamin.

One international unit is 1 mg alpha-tocopheryl acetate.

Chapter 35

Vitamin K

Vitamin K is used as a general name for a group of related substances widely distributed in plants, especially in dark-green leafy ones, and also in some animal products and micro-organisms. Common dietary sources are given in Table 31. In humans, as in other mammals, it is stored in the liver, with the other tissues having only very small amounts.

Functions

The most obvious function of vitamin K is to facilitate the production of the substances needed for the clotting of the blood. If healthy adults are fed less than about 20 μg vitamin K per day for a few weeks and if at the same time they are given antibiotics to sterilize their intestinal contents, the blood takes much longer to clot and may do so poorly. It is necessary to kill most of the intestinal micro-organisms because they normally produce vitamin K in the colon and enough of this can be absorbed into the blood to satisfy adult requirements.

In bone there is an abundant protein called osteocalcin and the production of this requires a good supply of vitamin K. Apart from binding calcium the function of osteocalcin is not fully understood. There are similar proteins in the kidneys.

Table 31 Vitamin K content of common foods

High (More than 100 μg/100 g edible portion)	Broccoli, cabbage, cauliflower, egg, lettuce, liver, spinach
Medium (50–100 μg/100 g edible portion)	Bacon, oats, watercress
Low (25–50 μg/100 g edible portion)	Butter, cheese, wheat (wholegrain)
Very low (Less than 25 μg/100 g edible portion)	Carrots, fruit, ham, maize, milk, peas, potatoes, tomatoes

Requirements

In the healthy adult there is need for only small amounts of vitamin K in the diet, or sometimes none at all, because the micro-organisms in the colon produce the vitamin. Dietary vitamin K becomes necessary, however, if the intestinal micro-organisms are killed by antibiotics and when there is appreciable chronic diarrhoea.

In the newborn there is always a very low level of vitamin K because the human placenta transfers the vitamin very poorly from mother to baby. In addition, the colon in the newborn is sterile and it takes several days for it to be colonized with vitamin K-forming micro-organisms. Breast milk and cows' milk contain only trivial amounts of the vitamin. As a result, one in a few hundred newborn suffers bleeding episodes during the first week of life due to vitamin K deficiency. To prevent this bleeding, newborn babies are given one dose of vitamin K to satisfy their needs to the time of weaning.

Because it is not known how much vitamin K is absorbed into the blood from the colon each day, recommended daily dietary intakes can be only speculative. It seems that about 1 μg of vitamin K per kilogram of body weight is sufficient. The average mixed diet in the United Kingdom provides at least about 300 μg per day, which is clearly enough. In the absence of disease there should be no need for dietary supplements of vitamin K.

Chapter 36

Folate

Folate, also known as folic acid or folacin, was once called vitamin M but this is no longer used. It derived its name from its abundance in dark-green foliage. Folate is a general name for a group of substances derived from pteroylmonoglutamic acid (PGA) by the addition of extra molecules of glutamic acid (an amino acid found in protein). In the early years of its study it was confused with vitamin B_{12} but this was clarified when folate was purified in 1943 and vitamin B_{12} purified in 1948.

Folate plays an essential part in the metabolism of rapidly dividing cells, such as those lining the small intestine and the cells in the bone marrow and spleen which produce the blood cells. It is needed for the synthesis of DNA, RNA and for many other syntheses including amino acids for protein formation and for the metabolism of long-chain fatty acids in the nervous system.

The normal body store of folate, about half of which is in the liver, is about 10–20 mg in a well-nourished adult. As the average daily need in a non-pregnant adult is about 0.3 mg, there is enough folate to last about two months on a deficient diet.

Sources

Folate occurs in nearly all natural foods of both plant and animal origin. The abundance in foods varies considerably and common sources are given in Table 32. Recorded values have increased in recent years because earlier ones were often under-estimated.

The amount of folate destroyed in cooking and processing food can be considerable, especially if the food is cooked twice. For example, mildly heating milk (up to about 75°C) may cause about half the folate to be lost; re-heating the milk may increase this loss to about three-quarters or even more. When heating food the folate loss can be minimized by keeping the food very slightly acidic; marked acidity rapidly damages the vitamin, as does making the food alkaline by adding sodium bicarbonate or similar material. Because of the easy destruction of dietary folate the most satisfactory way of obtaining the vitamin is from uncooked fruit and untoasted wholemeal bread. Generally, it is very uncertain

Table 32 Folate content of common foods

High (More than 100 μg/100 g edible portion)	Beans (kidney), bread (wholemeal), broccoli, Brussels sprouts, cabbage (dark-green), chicory, liver, peanuts, peas
Medium (50–100 μg/100 g edible portion)	Bananas, beans (runner), carrots, cauliflower, hazelnuts, orange, pork, potatoes, tomatoes
Low (Less than 50 μg/100 g edible portion)	Apples, beef, butter, cheese, chicken, cod, cornflakes (unfortified), egg, grapefruit, kidney, lamb, lettuce, milk, rice (white)

how much dietary folate is available for absorption into the blood unless the cooked food is analyzed, because the way in which food is prepared and kept is so variable.

Even very large doses (several hundred times the daily requirement) of folate seem to be non-toxic; the excess vitamin is excreted in the urine.

Requirements

Satisfactory daily intakes of folate for different groups of people are given in Table 33. These are the amounts of folate that should be in the diet each day and they take into account the fact that much of the vitamin will probably be lost in the preparation of the food. Non-pregnant adults need about 4 μg per kilogram of body weight, so that a 70 kg person will need about 300 μg. In the United Kingdom the average mixed diet contains 100–600 μg after the food has been cooked, so that some people may not get sufficient. During periods of increased metabolic activity, as in fever, infection, hyperthyroidism, pregnancy and lactation, the need for folate increases. Among medicines increasing folate need are the contraceptive pill and the anticonvulsants. Alcohol interferes with the absorption of folate and with its utilization and thereby increases the need for the vitamin.

Table 33 Folate satisfactory daily intakes

	μg
1–12 months	50
1–3 years	100
3–8 years	200
8–18 years	300
Adults	300
Pregnancy	800
Lactation	600

Deficiency

In the United Kingdom about 10 per cent of people older than 65 years have abnormally low levels of folate in the blood and a similar situation has been found in the United States. In both these countries about 20–25 per cent of untreated newly pregnant women are folate-deficient, showing abnormal changes in the bone marrow cells which produce the red blood cells. In very poor countries the occurrence of a megaloblastic anaemia due to folate lack is very common. This deficiency in pregnancy may cause damage to the developing baby and even miscarriage. The lack of folate is often made worse by lack of dietary iron, also common in pregnancy. A concomitant lack of vitamin B_{12} in pregnancy, which can also cause a megaloblastic anaemia, is unlikely because vitamin B_{12} deficiency almost always induces infertility.

A good supply of folate is essential in early pregnancy.

A study of folate deficiency produced by greatly over-cooking all food showed that it took about four months before there were signs of anaemia and that more than 100 μg of dietary folate were needed to cure the condition.

As well as producing a megaloblastic anaemia, folate lack causes lethargy, breathlessness and a smooth sore tongue. The condition is very similar to that seen in vitamin B_{12} deficiency and it is essential to make sure that the supply of vitamin B_{12} is adequate before giving folate. Increasing only the folate intake when vitamin B_{12} is needed may cause grave damage to the nervous system and it is imperative to seek specialist advice before treating any anaemia.

The importance of folate in pregnancy is described in Chapter 3.

Chapter 37

Niacin

Niacin is another name for nicotinic acid but it is also used in nutrition to indicate nicotinamide, which is the biochemically active form of the vitamin. It was at one time called vitamin B_3. It is a white crystalline material, stable to boiling and to light, air, acid and alkali, so that very little is lost during food preparation and processing. Although it was synthesized in the later part of the 1800s it was not recognized as a vitamin until 1937. It should not be confused with nicotine, which although a related compound has entirely different actions.

Niacin is unusual as it can be produced by the body.

Niacin is an unusual vitamin because, like vitamin D, it can be produced in the body. This is achieved by the conversion of the essential amino acid tryptophan to the vitamin. In the absence of adequate amounts of tryptophan in the diet, dietary niacin is necessary. When assessing the nutritional value of a diet both the niacin and the tryptophan should be measured and the results expressed as niacin-equivalents, although this is often not done.

Functions

Niacin (i.e. nicotinamide) is a constituent of enzymes involved in reactions releasing energy from protein, carbohydrate and fat. It is also involved in the synthesis of protein, fat and the pentoses (five-carbon sugars) needed for DNA and RNA. These reactions occur in all cells in the body but especially in the liver, which has a rich supply of the vitamin.

Sources

All foods contain some niacin although the amounts vary considerably. Liver is especially rich and so are peanuts. Most cereals have a useful niacin content except for maize which has only a small quantity of the vitamin and even this is not absorbable unless the maize is lime-treated, as for Mexican tortillas. In addition, the amino acid tryptophan is very low in maize protein (zein), so little of that is available for conversion to niacin. People living largely on maize are very likely to have a low niacin saturation.

Almost all the niacin (about 90 per cent) in cereals is in the outer husk so that it is lost if the cereal is milled to produce white flour and white rice.

About 100 g of animal protein (meat, poultry, fish, cheese) gives about 20 mg of niacin-equivalents from the tryptophan: this plus the niacin content of the diet will provide an ample supply of the vitamin.

Requirements

The tryptophan content of a diet is just as important as its niacin content; converted tryptophan can cure pellagra (niacin-deficiency disease) without the need for pre-formed niacin. In general, 60 mg of dietary tryptophan can produce 1 mg of niacin (1 niacin-equivalent). For most good quality proteins (of animal origin), about 1.5 per cent is tryptophan, and although not all this is available for conversion to niacin, there is enough to ensure health. For example, eggs, which contain little niacin, have enough tryptophan to prevent any niacin-deficiency illness.

The amount of niacin-equivalents needed is closely related to energy intake. About 11 mg of niacin-equivalents per 1000 kcal give a good margin of safety, so that for an average adult about 28 mg per day would suffice. Only about 11 mg of niacin-equivalents per day are required to prevent pellagra. If only pre-formed dietary niacin is being considered, then about 9 mg are needed each day.

Excess dietary niacin appears to be harmless unless massive doses are taken. The body store of the vitamin is small, with unneeded niacin being excreted in the urine.

Deficiency

> Niacin deficiency is rare in the United Kingdom except in alcoholics on low incomes or people with extreme food fads.

Lack of adequate dietary niacin together with a low intake of protein produces pellagra. This consists of a dermatitis in which the skin is very rough and dark in exposed areas, diarrhoea, dementia and an assortment of accompanying symptoms. Pellagra occurs in places where the diet is generally very poor and where maize is the main source of energy. This caused a serious epidemic in the southern United States from about 1865 to 1945, with tens of thousands of people a year becoming ill. Maize is a poor staple food because its niacin content cannot be absorbed without prior treatment with lime and because it is poor in tryptophan. In the United Kingdom pellagra is seen rarely and tends to occur in alcoholics on low incomes and in people with extreme food fads.

Chapter 38

Pantothenic acid and biotin

Pantothenic acid

Pantothenic acid was once called vitamin B_5 but this is no longer used. It is a yellow oil (isolated in 1938) which occurs widely in most foods and can be used as calcium pantothenate, which forms white crystals. It is stable at the usual cooking temperatures but being water-soluble some may be leached out when food is boiled.

Pantothenate plays an essential part in the metabolism of carbohydrate, protein and fat, especially in the reactions which release energy from these nutrients. It is also needed for the production of hormones, haemoglobin, cholesterol, chemicals which transmit impulses in the nervous system, antibody responses and the detoxification (neutralization) of toxins. For pantothenate to be properly utilized there must be an adequate intake of folate and biotin.

Sources

Because pantothenate occurs in all cells virtually any diet will provide a good supply of this vitamin. Everyday sources are given in Table 34.

Table 34 Pantothenic acid content of common foods

High (More than 5 mg/100 g edible portion)	Kidney, liver
Medium (1–5 mg/100 g edible portion)	Avocados, bread (wholemeal), cauliflower, egg, heart, herring, mackerel, mushrooms, oats, peanuts, soybeans
Low (Less than 1 mg/100 g edible portion)	Almonds, apples, bananas, beans (kidney), cabbage, carrots, grapefruit, grapes, kale, lettuce, milk, oranges, pears, plums, pork, potatoes, prawns, rice (white), spinach, sweet peppers, tomatoes, walnuts

Dry heating food, such as toasting, may greatly reduce the pantothenate activity and refining cereals for white bread and white rice usually causes loss of about half of the vitamin.

Requirements

Because pantothenate deficiency is very rare, the amount needed each day is not accurately known, but a safe intake for infants is probably 2–3 mg each day and for children and adults 5–10 mg. A mixed diet providing 3000 kcal per day will contain about 5–20 mg of pantothenate, depending on the foods chosen. Almost all of this will be absorbed into the blood, with about one-quarter being lost in the urine, leaving more than is likely to be needed for use by the tissues. Even for people on only about 1800 kcal per day the diet should contain enough of the vitamin.

Deficiency

A lack of pantothenic acid does not cause a clear-cut disease but results in a wide variety of vague symptoms because all the body tissues are to some extent affected. A lowered resistance to infection is often present with the immune system impaired. As the deficiency is most likely due to general malnutrition other vitamins are usually also missing and the symptom complex can be very confusing.

A diet inadequate in pantothenic acid in rats causes a greying of their hair but this is not the case in people. The use of pantothenic acid supplements will not prevent human hair from greying.

Biotin

Biotin, which forms colourless crystals, is part of the vitamin B complex; it is very stable to heat and light but because it is water-soluble some may be lost in cooking. It has also been called vitamin H. It occurs in virtually all foods and in addition it is synthesized by the micro-organisms in the human colon, from where variable amounts can be absorbed into the blood, so that people living on a diet devoid of biotin still excrete the vitamin in their urine. It may be that some people can aquire enough biotin from their colon to make dietary biotin unnecessary.

Biotin is involved in the synthesis of fat and protein and in the metabolism of carbohydrate.

Sources

Biotin occurs in small amounts in most foods and common sources are given in Table 35. Although eggs have a comparatively liberal amount of biotin, much of it cannot be absorbed into the blood because the white of egg contains avidin (a

Table 35 Biotin content of common foods

High (More than 200 μg/100 g edible portion)	Liver
Medium (30–60 μg/100 g edible portion)	Chocolate, egg (whole), egg (yolk), nuts, rice bran, soybeans
Low (Less than 30 μg/100 g edible portion)	Cauliflower, milk, mushrooms

protein) which forms an indigestible complex with biotin. One way to produce a biotin deficiency is to feed a large quantity of raw egg white (about 25 eggs are needed for an adult) and this method has been used in experiments. Rare cases of biotin deficiency occur in food faddists who consume many raw eggs each day. Cooking eggs changes the avidin so that it cannot bind to the biotin in the egg yolk or in the other items of the diet.

Requirements

The exact amount of biotin needed each day is not known because of the variable amounts absorbed into the blood from the colon, where it is produced by the micro-organisms. Estimates, however, have been made based on the amounts in the diets of healthy babies, children and adults.

Breast milk contains an average of about 1 μg of biotin per 100 ml, although the range is very wide; an infant getting 800 ml of breast milk per day has an intake of about 8 μg of biotin, which appears to be sufficient (the range is about 3–12 μg daily). The recommended intake, allowing a considerable safety margin, is 35–65 μg per day. For children the intake is set at 65–200 μg per day, according to age; while for adults it is 100–200 μg per day depending on body weight. It is assumed that about half these intakes will be absorbed into the blood. The average UK diet provides about 100–300 μg of biotin per day.

Biotin seems to have little or no toxic effects even on very high intakes.

Deficiency

Experiments have been carried out on people consuming excess raw egg-white for 10 weeks in order to prevent biotin absorption from the diet. They developed a scaly dermatitis, muscle pains, a smooth tongue, loss of appetite, nausea and an anaemia. They were cured by daily doses of 150–300 μg of biotin for 3–5 days, which would be helped by the biotin absorbed from the colon.

Chapter 39

Calcium, osteoporosis and phosphate

Calcium

Over 99 per cent of the adult body's 1000–1200 g calcium is in the bones and teeth, yet the remainder, less than 1 per cent, plays an essential part in the functioning of many diverse vital activities such as the transmission of nerve impulses, excitability at the junction of nerve and muscle, clotting of blood, absorption of vitamin B_{12}, maintenance and proper functioning of cell membranes, secretion of juices by cells, activation of enzymes and hormone secretion. The calcium in the blood is kept within a very narrow concentration range, using the bones as a source of calcium when needed and as a depot for calcium when the blood concentration is rising. This careful control is achieved mainly by the interplay of several hormones, a rise in calcium being brought about by parathormone from the parathyroid glands in the neck, while a decrease in calcium in the blood is brought about by the hormone calcitonin secreted by the thyroid gland. The responsiveness of bone to these hormones is affected by vitamin D, which is itself really a hormone even though it can be obtained in the diet. Many other hormones are also involved in the very elaborate system controlling the blood calcium concentration.

Peak bone mass is not achieved until 30 years of age and at 40, a decline in mass starts.

At birth, the bones are still soft but there is rapid calcification during infancy. At age six months the infant lays down in the bones almost as much calcium per day as does a 10 year-old (about 150 mg). At around 15 years there is a peak accumulation of about 400 mg of calcium per day, which then declines to around 50 mg per day at 20 years. Peak bone mass is not achieved until about 30 years of age. At about 40 years a steady decline in bone calcium starts so that by about 75 years the skeleton has only about 70 per cent of the calcium it had at its peak. This loss of calcium is more marked in women than in men, especially after the menopause. Bone fully calcified is about 50 per cent calcium phosphate, 20 per cent protein and the rest is made up of around 25 per cent water plus 5 per cent fat

Table 36 Approximate composition of the young adult skeleton

Total weight	9.00 kg
Minerals	4.50 (50% of skeleton)
Water	2.25 (25% of skeleton)
Protein	1.80 (20% of skeleton)
Fat	0.45 (5% of skeleton)

A young adult male skeleton has about 1.2 kg of calcium.

(Table 36). In poor dietary conditions there may be under-development of the skeleton because of protein lack and possibly also inadequate energy intake, such subjects being shorter and lighter than they would have been on a good diet.

The calcium in the skeleton is not static but is constantly renewed.

The calcium in the skeleton, unlike that in the teeth, is not static but is constantly being renewed. In an adult this calcium renewal (turnover) amounts to about 500 mg per day, so that all the calcium in the skeleton is renewed about once in seven years. The turnover is very much faster in young children, taking as little as one year to renew all the calcium. As well as the interplay of many hormones and other substances controlling the calcification of bone, the amount of calcium laid down is greatly increased by exercise, especially when lifting and carrying are involved. Conversely, prolonged bed rest causes a rapid loss of bone calcium. It is very important for the skeleton to accumulate as much calcium as possible during the early years because fully calcified bones take longer to develop osteoporosis than do less dense bones.

Lifting and carrying are very important for good calcification of bones.

As well as calcium, bone accumulates related minerals such as strontium, caesium and lead. If lead is ingested, usually in the water, it will be laid down in bones and will be released gradually later, thereby producing a state of chronic lead toxicity. The most likely source of lead is from piping, especially in soft water areas. Strontium and caesium were dangerous in the 1950s and 1960s because their radioactive forms were released into the atmosphere from nuclear weapons tests.

Absorption of calcium

Dietary calcium is absorbed into the blood mainly by the first 30 cm (1 ft) of the small intestine, where the lining cells are specially adapted for this purpose. If this area is missing or damaged by disease there will be inadequate calcium uptake from the normal diet. The absorption of calcium by the normal adult is not particularly efficient, being only 30–60 per cent when the intake is about 400–1000 mg calcium per day, which is less than desirable, although it was thought to be enough until the 1980s. Even this low absorption rate gets worse as the calcium intake increases. In actively growing children, in contrast, the absorption rate can be as much as 75 per cent of intake.

The efficiency of calcium absorption is affected by vitamin D, which increases uptake, and by other dietary constituents, which generally reduce the uptake. In ordinary UK foods the chief items are phytate and oxalate. Phytate occurs mainly in cereals, with oats containing the greatest amount. This substance is mostly in the outer layer of the grain, so that wholemeal bread and wholegrain rice contain more of it than do white bread and white rice, from which the outer layers of the grain are missing. This is why on first changing to a wholemeal diet there may be a sharp fall in calcium absorption but after a while the intestine produces more of an enzyme (phytase) which breaks down the phytate and thereby permits the calcium uptake to rise. Nevertheless, when wholemeal foods constitute a large part of the day's energy supply, extra calcium-rich food and vitamin D are desirable to keep the body's calcium supply sufficient. The phytate in 250 g of wholemeal bread (about seven slices) can prevent up to 300 mg of calcium from being absorbed, which is likely to be about one-third of an average daily intake. Oxalate occurs in much smaller amounts than does phytate in ordinary food in the United Kingdom, the main sources being spinach, rhubarb and soybean, in which there is often enough oxalate to prevent any of the calcium in those items from being taken up by the intestine. It is unlikely, however, that there will be enough oxalate to affect the calcium of other foods eaten at the same time as the spinach, rhubarb or soybean.

Phytate and oxalate reduce calcium absorption.

For calcium to be absorbed efficiently, the contents of the upper part of the small intestine must be acidic, which they are normally. In old age this acidity often declines, resulting in a lowering of calcium absorption.

Sources

In the everyday UK diet the only foods rich in calcium are milk and milk products. Not only is milk rich in calcium but the mineral is very well absorbed from milk and milk is cheap. If milk and its products are not eaten it is almost impossible to get a good calcium intake unless the diet is supplemented with calcium tablets. One litre of cows' milk (1.7 pints) contains about 1200 mg of calcium (one pint has about 700 mg). The removal of

There are about 700 mg of calcium in one pint (0.6 litre) of milk and in 100 g (3.5 ounces) of hard cheese.

Table 37 Calcium content of common foods

Very high (More than 500 mg/100 g edible portion)	Hard cheese
High (200–300 mg/100 g edible portion)	Almonds, chocolate (milk), hazelnuts, kale, soybeans, Stilton cheese
Medium (50–200 mg/100 g edible portion)	Beans, Brazil nuts, bread (unfortified), broccoli, celery, chicory, chocolate (plain), cottage cheese, cream, dates (dried), egg, figs (dried), lentils, lobster, milk, olives (green), oyster, parsnips, peanuts, pecans, prawns, prunes, raisins, raspberries, walnuts, watercress, yogurt
Low (Less than 50 mg/100 g edible portion)	Barley, Brussels sprouts, butter, cabbage, carrots, cashews, cauliflower, cornflakes (unfortified), fish, fruit, lettuce, meat, peas, potatoes, poultry, rice, sweet peppers, tomatoes

the fat from milk leaves behind all the calcium so that semi-skimmed and skimmed milk are just as calcium-rich as is full-fat milk. For hard cheese, 100 g (3.5 ounces) has about 600–700 mg of calcium, which is much the same as one pint of milk.

For children, only full-fat milk should be used below the age of two years. From 2–5 years semi-skimmed milk may be used if the child is overweight. Over five years of age skimmed milk may be given if there is a need to control the energy intake.

Fresh hard water may contain enough calcium to provide 20 per cent of the daily requirements.

The calcium content of everyday foods is given in Table 37. Although soft water has almost no calcium, fresh hard water may contain sufficient to provide about 20 per cent of the daily need; boiling hard water causes almost all the calcium to be precipitated as an insoluble material. In communities where milk is unavailable, much of the dietary calcium may come from the bones of animals, especially fish. There is little or no danger of producing kidney stones in a normal person by the amount of calcium that can be absorbed from ordinary food. The intake is unlikely to exceed 2000 mg per day, which is a safe level.

Requirements

Dietary surveys in western countries show that the daily calcium intake has a very wide range of 400–1300 mg, with men consuming more than women. In

Table 38 Calcium satisfactory daily intakes

	mg
0–3 months	500 (or as in breast milk)
3–12 months	500–800
1–8 years	800
8–20 years	1200
20–60 years	1000–1500
Over 60 years	1500
Pregnancy	1200–1500
Lactation	1500–2000

the United Kingdom in 1990 the daily value for non-pregnant women was about 500 mg, while in large parts of Africa, Asia and South America it was even lower at around 300 mg of calcium per day. Even in the United States in 1980 only about one-third of young women had the recommended daily calcium intake. Most of these people seem to adapt by improving their intestinal absorption efficiency so that calcium loss in the faeces is greatly diminished. In many the defects that might be expected from such calcium deprivation are not prevalent even during pregnancy and lactation.

The daily calcium intakes recommended in different countries vary considerably, especially since about 1990, when values were generally greatly increased. Most of the changes in recommended intakes have been based on recent research into osteoporosis (brittle bone disease) and the realization that the best preventive measure is to have maximum bone calcification in childhood and in the early adult years. Table 38 gives daily values which are generous and probably at about the maximum that can be achieved on an ordinary UK diet without supplementation by calcium in tablet form. Currently only a minority of the population reachess these intake levels.

Many western diets have a liberal amount of phosphate, which lowers the calcium to phosphate ratio thereby causing an increased loss of calcium in the urine. Much of this phosphate is derived from protein, so that a diet rich in protein increases the need for calcium.

After many years on a diet rich in calcium, adaptation to a low calcium intake can be very slow and may not reach a satisfactory level. This is very important for those, especially older women, changing to a slimming diet; they must keep their calcium intake high, taking calcium tablets if necessary.

Osteoporosis

Osteoporosis (brittle bone disease) is the name given to the condition in which bones become brittle, but not soft, because calcium salts are lost from their substance. The bones do not shrink but remain their original size. The skeleton

Bone loss occurs in both sexes but is four times greater in the female after the menopause.

achieves its peak mass at around 30 years of age and remains stable for about 10 years after which it begins to lose mass because the bone which is regularly removed throughout life is not completely replaced: the turnover is in negative balance. This age-related bone loss occurs in both sexes but is about four times greater in the female after the menopause and reflects the falling level of the hormone oestrogen. Unlike men, who do not usually have sufficient loss of bone calcium to cause inconvenience, women by the age of 70 years will have lost an average of about 50 per cent of their bone mass and about a quarter of all women of that age will have had at least one bone fracture. The parts chiefly affected are the vertebrae, the top end of the femur (the thigh bone) and the wrist. The compression of the brittle vertebrae leads to marked loss of height and the bowing of the upper part of the back seen in many older women. It is often accompanied by pain and sometimes by problems related to trapped nerves. Small-boned women who have had a sedentary life are most at risk, especially if they have had repeated spells of dieting. The bones heal slowly and less well and immobility during treatment makes bone loss worse. Any condition which reduces oestrogen levels in young women will help to produce osteoporosis, so that anorexia nervosa and excessive exercise, such as marathon running, which lead to amenorrhoea, are two well-known causes. Some types of exercise, however, are very valuable in preventing bone loss and they are those which involve lifting and carrying, the load-bearing bones laying down extra calcium as a result of the strain placed upon them.

Lifting and carrying help prevent osteoporosis.

As with many conditions, prevention is far more satisfactory than cure. Treatment of osteoporosis never brings the bones back to their optimum state even though some benefit can often be obtained. The best prevention is secured during childhood and young womanhood by a diet rich in calcium, protein and the vitamins, especially vitamin D produced by sunlight on the skin or taken in the diet, usually in fortified foods. The daily calcium intake should be about 1500 mg, which can be achieved by taking 600 ml (one pint) of skimmed milk per day (700 mg of calcium) plus about 100 g (3–4 ounces) of low-fat hard cheese (700 mg of calcium). If this is not feasible, calcium tablets should be used. Total daily calcium intake need not exceed 2000 mg as consumption of more than about 1500 mg per day does not confer greater protection. The daily vitamin D in the diet should be 2.5–10 μg if there is less than 1–2 hours sunning of the skin. Added to these dietary measures there should be regular exercise involving lifting and carrying. When bringing shopping home it is better for the woman to carry it rather than the man. Any abnormality which may limit calcium absorption must be adequately treated. Despite these measures, some

Table 39 A vegan diet supplying about 1000 mg of calcium per day

	mg calcium
90 g (about 3 ounces) soybean curd	425
600 ml (about one pint) soya milk	75
Vegetable and bean casserole using soya cheese	260
4 large slices fortified wholegrain bread	120
60 g (about 2 ounces) muesli	70
30 g (about 1 ounce) dried figs	80
90 g (about 3 ounces) peanuts	50

women seem likely to endure some degree of ostoeporosis because peak bone mass is primarily genetically determined: girls with a family history of osteoporosis should take preventive measures early and seriously. There is evidence that cigarette smoking and alcohol may make osteoporosis worse.

A vegan diet containing about 1000 mg of calcium is given in Table 39.

Hormone Replacement Therapy (HRT)

Hormone Replacement Therapy (HRT) is the name given to the treatment of osteoporosis by supplements of the female hormone oestrogen. It is the lack of this hormone after the menopause which greatly accelerates loss of bone calcium and therefore HRT has its most beneficial effect when started at the menopause. After osteoporosis has occurred HRT is less valuable but still exerts a worthwhile effect. Its greatest action seems to be stopping calcium loss from the vertebrae rather than from the limb bones. It also has a beneficial effect on calcium absorption from the intestine because it increases the synthesis of vitamin D by the kidney. If the uterus is still present a second hormone, progestogen, is also given to reduce to normal the risk of uterine cancer. HRT is usually not given for longer than ten years.

For women over 60 years of age, for whom HRT is not an option, a calcium intake of around 1500 mg per day plus 2.5–10 μg of vitamin D plus daily moderate physical activity may slow the rate of bone calcium loss. If there is already present appreciable osteoporosis the exercise routine must be very carefully graded so as not to cause a bone fracture.

Osteoporosis in men

Vertebral fractures and hip fractures in older men are only about one-fifth as common as in older women. This is partly because most men continue to produce male sex hormones in liberal amounts well into old age and partly because men develop denser bones in their earlier years and can therefore tolerate some calcium loss. Over the age of about 80 years the rate of hip fracture becomes similar in men and women because by then the secretion of male sex hormones has fallen to a very low level.

> In communities where women do most of the physical work, men have more osteoporosis than the women.

As in women, men taking diets rich in calcium and with adequate vitamin D produce dense bones during the first 20–30 years of life and this is aided by physical effort involving lifting and carrying. Inactivity, smoking and excessive alcohol have a deleterious effect on bone calcification. In communities in which women do most of the physical work the men have more osteoporosis than do the women. The dietary advice for older men is the same as that for older women.

Osteomalacia

> Rickets in the young and osteomalacia (soft bones) in later life are prevented by vitamin D.

This is the name given to the condition in which bones remain or become soft. They are not brittle. It is mainly due to a lack of vitamin D over a long period although there are also other causes. Because the bones are soft they bend, producing the noticeable bowing of the legs, but more important for women is the narrowing of the pelvic outlet, sometimes producing grave difficulties at childbirth. Because mis-shaped bones cannot be cured easily, it is very important to prevent this disease, rickets, from occurring by ensuring that from infancy onwards the body has enough vitamin D, either from sunlight on the skin or from vitamin D-fortified food. This is of especial concern for girls from cultures demanding that the whole skin be covered. It is essential for them to eat food enriched with vitamin D, taking up to 10 μg (micrograms) per day, which is a safe amount. The treatment of rickets should be under the care of a specialist.

Phosphate

> Phosphate is essential in all reactions in the body that release energy.

About 1 per cent of the adult body is phosphorus, amounting to about 700 g, of which 85–90 per cent is in the bones and teeth, about 5 per cent in the muscles and the rest in the remaining tissues. It is virtually always present as phosphate and is found in every cell, being part of DNA, RNA, the cell membranes and many enzymes. It is essential in all reactions releasing energy, it helps to control the alkalinity of the blood and it takes part in virtually all cellular reactions.

The intake of phosphate by the adult should be about equal to that of calcium. In the first six months of infancy, the ratio of calcium to phosphate should be about 1:8, as it is in breast milk. Almost all ordinary UK diets have a calcium to phosphate ratio of 1:2 and there is little chance of there being enough phosphate to cause a problem.

Sources

Phosphate is found in virtually all foods and is high in those rich in protein such as meat, poultry, fish, eggs and cereals. There is no phosphate in sugars, fats and oils. The average daily intake on a mixed diet in the United Kingdom is about 1.5 g. The absorption into the blood of phosphate is uncontrolled, with the kidneys excreting any excess. Under ordinary conditions about two-thirds of dietary phosphate is excreted in the urine and the rest is in the faeces, shed skin and shed hair.

Cows' milk has a calcium to phosphate ratio of 1:4, lower than that for human milk, and if fed to infants during the first few weeks of life may limit the absorption of calcium, producing tetany, a condition in which the muscles go into spasm (milk tetany). Unmodified cows' milk should not be fed to infants for this and other important reasons.

Dietary lack of phosphate is exceptionally rare and a low level of phosphate in the blood is likely to occur only in kidney disease or as a result of renal dialysis. Premature infants need more phosphate than there is in breast milk and may show signs of deficiency if this is not taken into account.

Chapter 40

Iron

About two-thirds (3 g) of the iron in an adult is carried in the red blood cells (erythrocytes) in a red-coloured molecule called haemoglobin. This substance carries oxygen from the lungs to the tissues, carries carbon dioxide from the tissues to the lungs and helps to keep the blood at the correct degree of alkalinity. The haemoglobin is inside the red cells to prevent it being excreted by the kidneys. The rest of the body's iron (1.5 g) is in enzymes, in co-factures, in muscle myoglobin and in storage as ferritin and haemosiderin mainly in the liver, the spleen and the bone marrow. The amount in store is usually 0.5–1.5 g but this can vary considerably even in health. Almost all the stored iron comes from cells, including red blood cells, which are broken down and their iron content re-used. About 20–25 mg of iron is needed each day for the synthesis of new haemoglobin but only 1–2 mg of this is newly absorbed dietary iron, the rest being iron which has been previously used and recycled. The body is exceptionally efficient in its iron usage. In health, the urine is virtually iron-free and the iron in the faeces is always less than the dietary iron. Iron is lost from the body only by the rubbing off of skin and hair and by bleeding. Under normal conditions the adult man and the non-pregnant non-menstruating woman lose only about 1 mg of iron a day. As the store in health is 500–1500 mg it would normally take several years for the store to be depleted on a diet deficient in iron.

Anaemia due to lack of dietary iron usually takes several years to develop.

Sources and requirements

On an ordinary UK diet, iron is derived about one-third from animal foods, one-third from fortified breakfast cereals and one-third from the rest of the diet. The iron content of common foods is given in Table 40 and the daily iron needs in Table 41. On average, 10–15 mg are consumed each day and the efficiency of absorption is around 10 per cent. Most men get enough dietary iron but many women, especially if menstruating or pregnant, do not.

Table 40 Iron content of common foods

High (More than 6 mg/100 g edible portion)	Clams, heart, kidney, kidney beans, liver, lentils, oyster, soybean, tongue
Medium (4–6 mg/100 g edible portion)	Almonds, cashews, figs (dried), hazelnuts, prunes, rice (wholegrain)
Low (2–4 mg/100 g edible portion)	Bacon, barley, beef, Brazil nuts, chicken, chocolate, egg, ham, kale, lamb, peanuts, peas, pecans, pork, rabbit, raisins, turkey, veal, walnuts, watercress, wheat (wholegrain)
Very low (Less than 2 mg/100 g edible portion)	Cheese, cornflakes (unfortified), fish, fruit, milk, prawns

Table 41 Satisfactory daily intakes of iron

	mg
Infants	5–15
Children	5–10
Adolescent boys (50–100 kg)	10–20
*Adolescent girls (45–75 kg)	10–25
Adult males (65–100 kg)	10–15
Non-menstruating women (50–75 kg)	5–10
*Menstruating women (50–75 kg)	10–25
*Pregnancy (50–80 kg)	15–35

*Will often need supplements, especially if on a diet or eating little meat.
Upper part of range needed for heavier subjects.

Vitamin C aids the absorption of iron.

The iron in some foods is only poorly, or very poorly, absorbed into the blood by the intestine. In general, iron in animal products is well absorbed while plant iron is not. An exception to this is egg, from which only about 2 per cent of the iron can be used because of the high phosphate content in egg. In contrast, about 20 per cent of the iron in steak, pork, lamb, poultry and fish enters the blood. An example of a plant with a good iron content which is not very useful is spinach, because the large amount of oxalic acid in spinach prevents much of the iron from being absorbed. On the other hand, 20 per cent of the iron in soybean can be absorbed into the blood, which is exceptional for a plant food. For all foods, the presence of vitamin C aids the absorption of iron so that vitamin C-rich food, such as orange juice, is best taken with meals.

Iron absorption is much more efficient in the young than in the adult and it increases when there is a need for iron, as in certain types of anaemia and during pregnancy, when up to 30 per cent of the iron in animal food and 20 per cent of the iron in plant food can be absorbed. The control of iron absorption into the blood is exerted by the cells lining the upper part of the small intestine. As this controlling mechanism can be overcome by an excessive intake of iron, supplements of iron should not be taken except on medical advice. Iron overdose can cause severe illness.

A vegetarian diet may contain what seems to be an adequate amount of iron but if the dietary fibre in the food is very high the amount of iron absorbed into the blood may be too low. The inclusion of eggs and milk products in a vegetarian diet does not help because the iron of eggs is very poorly absorbed and milk has only a trivial amount of iron. Many vegetarian women of childbearing age have a low haemoglobin concentration in their blood. A breakfast cereal well-fortified with iron should be taken each day.

Iron in infancy

Healthy newborn babies have an iron reserve in their livers.

Healthy newborn babies have an iron reserve in the liver which satisfies their needs for 3–6 months. This is essential because milk contains little iron and for the first three months the immature intestine can absorb only very small amounts of the metal. Iron from broken-down red blood cells, which are especially abundant in the newborn, is carefully re-used.

After the first three months, because there is a marked increase in iron usage for the rapidly growing tissues and the expanding blood volume and because it is not possible to be sure that the iron stores are sufficient, iron-enriched food should be given and continued until the child is at least 1 year-old. Generally, an intake of about 1 mg of iron per kilogram of body weight per day is adequate during the first year.

Anaemia during infancy is usually due to a low reserve of iron in the liver at birth because of iron-deficiency in the mother. It is more likely to occur in infants with a low birth weight, in premature infants and in twins.

Unmodified cows' milk should not be fed to infants because in about one-third of them it causes bleeding into the intestine.

Menstruation

The amount of blood lost during each menstruation varies considerably from woman to woman but is usually constant in each individual over several years. Losses range from as little as 5 ml of blood over a 28-day cycle to as much as 160 ml. This represents about 0.1–3 mg of iron per day, with the average being about 0.7 mg per day. Losses tend to rise with advancing years. They are increased when an intrauterine contraceptive device is used and are decreased when contraceptive pills are used. Most women with heavier losses are likely to be permanently mildly anaemic unless they take a diet rich in iron and protein.

Iron deficiency

Iron deficiency is one of the commonest nutritional deficiencies, not only in poor countries but also in the richer ones. It affects chiefly women and children. In the United Kingdom it is particularly prevalent among Asian people in the large cities. Women who have had repeated pregnancies and women with heavy periods (menorrhagia) are often permanently mildly anaemic and so is anybody who has chronic marked bleeding, whatever the cause.

Most often iron deficiency arises because of an iron-poor diet but it can also be due to a defect in intestinal absorption of iron. Because the store of iron in a normally well-fed person is large, it can take several years for an iron-deficiency anaemia to develop when changing to an iron-poor diet.

The severity of the symptoms in anaemia vary considerably from person to person. They may include weakness, tiredness, loss of appetite, breathlessness on exercise, palpitations, sore tongue and swollen ankles. Because the symptoms are non-specific a correct diagnosis may be delayed.

Before iron supplements are taken for anaemia it is essential for the cause to be diagnosed by a specialist.

Chapter 41

Sodium, potassium and chloride

Sodium

There are about 100 g of sodium in the adult, about 10 per cent being inside the cells, about 50 per cent in the fluids outside the cells (including the blood) and about 40 per cent in the bones, which act as a reserve from which sodium can be released to keep the blood level constant. It plays an essential part in controlling osmotic pressure, in the conduction of nerve impulses, in muscle contraction and in the transport of substances into and out of cells.

Sources and requirements

In a temperate climate as little as 0.25 g of dietary sodium per day is probably enough, although most people need about 1 g per day to feel comfortable. In the United Kingdom the intake has a range of about 2–10 g of sodium, which is equivalent to 5–25 g of table salt (sodium chloride). This daily intake comes from about 2 g of salt added at the table plus the sodium in the food and added during food preparation. As sodium is present in virtually all food it is difficult to take less than about 1 g per day. On the other hand, taking relatively liberal amounts of sodium is commonplace: popular foods high in salt are shown in Table 42. For some foods salt is added for its taste, while for others it is added as a help in preservation.

The taste for salt is acquired during infancy when food seasoned with salt is fed. Infants on weaning do not discriminate between salted and unsalted food but they gradually acquire a liking for salt which is difficult to break.

To reduce sodium intake food should be thoroughly cooked in plain water. Such food is often not very appetising but can be improved by the addition of spices, which generally contain very little sodium.

Tap water in the United Kingdom sometimes has a relatively high sodium content and some water-softeners produce water high in sodium.

Table 42 Some popular foods often high in salt

Bacon, baked beans, canned meats, canned soups, cheese (hard), corned beef, cornflakes, cottage cheese, crisps (salt added), fish fingers, ham, kippers, olives, peanuts (salted), pizza, sausages, smoked haddock, tomato ketchup

Sweat has an appreciable sodium content and when sweating is profuse it may be beneficial to add a little extra table salt to food. The use of salt tablets is rarely necessary and if they are used they should also contain a supplement of potassium.

Deficiency

Sodium deficiency is very rarely of dietary origin, although it may occur during starvation. It can be induced by excessive water intake, the sodium being lost in the large output of urine. Sodium can also be lost in considerable amounts in severe vomiting, diarrhoea, renal failure, adrenal gland failure, excessive use of diuretics, chronic wasting diseases, extensive burns, major surgery, severe injury and in excessive sweating. The results of a severely lowered blood sodium concentration are nausea, anorexia, muscle weakness and spasm, headache, confusion, coma and death. Treatment of any of these complications needs expert advice.

High blood pressure (hypertension)

In some people, perhaps about 10 per cent of the population, high blood pressure is lowered by a reduction in sodium intake. The systolic pressure responds more than does the diastolic pressure and the response is greater in older people with relatively high pressures. The giving-up of salt at the table with only moderate use of salt in cooking sometimes induces a worthwhile fall in the blood pressure and may be preferable to the taking of anti-hypertension drugs.

Potassium

Potassium is an essential component inside every cell, where it is involved in many vital reactions. Among these are the maintenance of the correct osmotic pressure, the conduction of impulses along nerve fibres, the rhythmic contraction of the heart, protein synthesis, carbohydrate metabolism with the release of energy, the release of the hormone insulin by the pancreas and the maintenance of a normal blood pressure.

Sources and requirements

The ordinary UK diet provides more potassium than is needed each day and the excess absorbed is normally excreted in the urine. Very good sources are oranges,

The ordinary UK diet provides more potassium than is needed each day.

bananas, meat, haddock, cauliflower, potatoes, Brussels sprouts, broccoli and tomatoes. The daily dietary intake in the United Kingdom is about 2–4 g, with vegetarians having especially large amounts. In a fully-grown adult the amount absorbed into the blood each day is balanced by the amount lost in the urine. When new tissue is being formed during growth some potassium is retained for use in the new cells. The minimum needed each day is uncertain but is probably about 2 g for healthy adults.

Deficiency

Potassium deficiency is rare in the United Kingdom in the absence of diabetes, severe diarrhoea, repeated vomiting, adrenal gland disease, severe damage to tissues (especially muscle), surgery, some diuretics and excessive use of laxatives. A marked fall in body potassium may produce muscular weakness progressing to paralysis, a rapid and irregulr heartbeat progressing to sudden cardiac arrest and failure of adequate breathing because of muscular weakness. If tablets are taken to relieve potassium deficiency (hypokalaemia) care is needed because high concentrations of potassium are very irritant to the intestine.

Potassium excess

Excessive blood levels of potassium are very unlikely to occur unless there is kidney failure, adrenal gland disease or very severe damage to tissues causing them to release their potassium into the blood. It may, of course, be caused by excessive supplementation with potassium tablets. The results are very similar to those seen in potassium deficiency, namely muscle weakness, paralysis and cardiac arrest.

Chloride

Chloride is abundant in virtually all normal foods and dietary lack is unknown. Sufficient can, however, be lost from the body to cause severe symtoms during prolonged profuse sweating, repeated vomiting and intractable diarrhoea. The chloride ion, together with sodium, plays an essential role in maintaining fluid balance in the body and in the absence of enough chloride there is loss of body water via the kidneys, resulting eventually in collapse and death because of inadequate functioning of the cardiovascular system.

As well as controlling body water, chloride is the main negative ion and is needed to balance the positively charged ions in all the body tissues, the two being inextricably linked. Hence loss of much sodium or potassium in the urine will automatically result in the loss of chloride. Many diseases may upset this normally well-controlled complex relationship but dietary involvement is rare.

There is so much chloride in the UK diet that it is very difficult to produce a chloride-deficient diet that is palatable. The daily salt (sodium chloride) intake varies widely, being on average about 15 g per day, of which about 6 g will be chloride. In tropical regions it may reach 50 g of salt per day.

Recommended intakes at various ages are shown in Table 43, but these values have little relevance to the daily diet.

Table 43 Chloride satisfactory daily intakes

Age (in years)	*mg*
0–1	300–1200
1–6	500–2000
6–10	900–3000
10–18	1400–4000
Over 18	1700–5000

Chapter 42

Iodine

> If iodine intake is insufficient the thyroid gland swells and produces a smooth, painless lump called a goitre.

Iodine is essential for the production of the hormones of the thyroid gland, which lies alongside the larynx in the front of the neck. In the absence of an adequate iodine intake, the thyroid gland enlarges to produce a smooth, painless, rounded lump, easily seen moving up and down during swallowing. In many parts of the world, especially in rainy upland areas, iodine is in short supply and enlarged thyroids, called goitres, have until the early 1900s been common, particularly in young women. Goitres have been known since ancient times and were treated with burnt seaweed and sea sponges, which are rich in iodine, although the reason for the success of this therapy was not understood until the early 1800s when iodine itself was found to cure simple goitre. At the end of the 1800s iodine was shown to be concentrated by the thyroid gland and in the early 1900s it was identified in the thyroid gland hormone called thyroxine. In the middle 1800s iodine was actually used in France to treat goitre in school children but because too much was given, producing toxic symptoms, the treatment lapsed for about fifty years, when its use became widespread in goitrous regions. There are, of course, many other causes of goitre, apart from lack of iodine.

Dietary iodine is easily and almost completely absorbed but there is no body store of the element, so that a regular frequent intake is necessary. Excess iodine is excreted in the urine.

The thyroid hormones increase the rate of metabolism of all cells, sometimes by as much as 30 per cent. The hormonal activity is therefore closely related to the metabolic rate of the whole body, which can be measured by the rate at which oxygen is taken up. In the unborn baby, thyroid hormones are essential for proper physical and mental development.

Sources

The amount of iodine in a food depends on the place in which the food was grown. Where the soil is rich in iodine, plants have a high iodine content and the tissues of animals eating such plants also become iodine-rich. For example, iodine in milk has been found to have a range of about 20–300 μg per litre, depending on where the cows were living. Since the late 1900s food in the United Kingdom has been available from widely different regions, rather than from a single region, which evens out marked differences in food iodine content.

Iodized table salt provides a valuable source of iodine.

The use of iodized salt, containing about 2 mg of potassium iodide per 100 g of salt, is a simple and very useful nutritional measure and costs very little. A person eating 2 g of such a fortified salt each day would get 30 μg iodine, a very valuable addition to the daily iodine intake. The amount of iodine added to salt is never enough to be toxic.

Although seawater contains iodine, sea-salt is not a good source of iodine because most of it volatilizes when the seawater is dried. Salt-water (sea) fish, particularly haddock, is a useful common source of iodine.

Requirements

The recommended iodine intake for children and adults is about 100–150 μg per day, although a range of 50–1000 μg of iodine per day is normally safe for adults. An average UK diet when using iodized salt provides about 500 μg iodine per day.

Iodine needs in pregnancy and lactation are dealt with in Chapter 3.

Intakes of iodine of 1000 μg per day or more have little effect on a normal thyroid gland. The gland takes up more iodine than usual for a few weeks and is then back to normal. Hormone release is usually unchanged, or may be trivially reduced at first. Excess iodine intake can occur as a result of undue consumption of some seaweed products.

Chapter 43

Fluoride

Fluoride is particularly important because it is the only nutrient to substantially reduce dental caries. In the early 1900s it was observed that caries was lower when the drinking-water contained about 2.5 parts fluoride per million (2.5 mg per litre) than when the content was less than about 0.3 parts per million. Subsequent work showed that adding fluoride to the water was as effective as was the naturally occurring fluoride. During these trials the number of decayed teeth in the fluoride-enriched areas fell by up to 70 per cent. For the best effect on teeth, a water fluoride level of 1 part per million is required. The fluoride makes the enamel of the teeth more resistant to bacterial acid and it reduces the amount of bacterial acid produced. In the United Kingdom many children are still deprived of fluoride-enriched drinking-water even though fluoridated water has been supplied to many millions of children and adults over several decades without any evidence of harm.

Adding fluoride to drinking-water can reduce dental caries by up to 70 per cent.

Sources

The main source of dietary fluoride is drinking-water except in areas where the fluoride level is especially low (below 0.3 parts per million). Tea is a rich source, giving about 0.1 mg of fluoride in a cup of average strength. Seafood in general has a good fluoride level (5–15 parts per million), with mackerel high at around 30 parts per million. Vegetables, wholegrain cereals and fruit are very variable. An ordinary mixed diet in the United Kingdom contains about 1–2 mg of fluoride per day. Breast milk is low in fluoride, so that in low-fluoride areas supplements need to be given to the newborn.

Fluoride supplements

In areas where the drinking-water has more than 0.7 parts fluoride per million, supplements should not be used. For areas with less than that the supplementation

Table 44 Fluoride supplementation using tablets containing 0.55 mg of sodium fluoride

a) When drinking-water has less than 0.3 p.p.m. fluoride

Years	*Tablets per day*
Under 2	1
2–4	2
4–16	4
Over 16	0

b) When drinking-water has 0.3–0.7 p.p.m. fluoride

Years	*Tablets per day*
Under 2	0
2–4	1
4–16	2
Over 16	0

When drinking-water has more than 0.7 p.p.m. fluoride, no supplements should be used.

details are given in Table 44. Supplements are supplied either as drops, which are added to drinks, or as tablets, which should be allowed to dissolve in the mouth and not swallowed whole. Children using these supplements have on average only 20–25 per cent of the caries found in children who are not so treated.

As soon as teeth erupt they should be cleaned with a fluoride toothpaste provided the child learns not to swallow the toothpaste. A piece of toothpaste the size of a garden pea is all that need be used. In addition to toothpaste, mouth rinses containing fluoride can be used daily and special fluoride pastes can be applied to the teeth by a dentist at six-monthly intervals.

Toxicity

When drinking-water has more fluoride than about 2.5 parts per million the tooth enamel may become mottled (chronic endemic dental fluorosis), although the teeth usually remain sound and highly resistant to caries. The higher the fluoride level, the greater is the mottling. In areas of particularly high fluoride levels, greater than about 8–10 parts per million, the teeth may become chalky and erode rapidly. When the fluoride level reaches about 15 parts per million, as it does in some parts of the world, fluoride poisoning occurs, a feature of which is a very stiff back due to calcification of spinal ligaments. In such places it is necessary to remove the fluoride from the water.

Fluoride supplements must not exceed the recommended dosage.

Chapter 44

Selenium

> Selenium is required for normal testicular function.

For several years it was known that there was a relationship between vitamin E and selenium in that they could partially replace each other. They both prevent unwanted oxidation of fat within the cells and in the cell membranes, they reduce the occurrence of atherosclerosis (fatty damage to the arterial walls), protect DNA and inactivate carcinogens. In selenium lack, the thyroid hormone tri-iodothyronine is reduced and growth hormone production falls. Selenium is also involved in the reactions which release energy in cells. It is necessary for normal testicular function and for the normal development of sperm cells. It performs these functions by being part of a group of compounds called seleno-proteins, which are essential components of a range of intracellular enzymes. Most of the actions of selenium were discovered between about 1950–1975.

Selenium is most abundant in plants grown in selenium-rich soil, as in the United States, whereas most European-grown plants are poor in selenium, unless it is added to the soil, which is now done in most countries. This addition of selenium must be carefully gauged because excess can result in toxicity to farm animals.

Good sources of selenium are wholegrain cereals, offal, fish and meat, if they come from a selenium-rich environment. Brazil nuts are very rich in selenium. Milk and milk products are low in selenium unless the element is added to the animals' food, which is now usual in the United Kingdom.

Requirements

The amounts believed to be needed each day for various groups are given in Table 45. The UK daily intake fell from about 60 μg of selenium per day in 1974 to about 34 μg per day in 1994 because less United States-grown food and more European-grown food was eaten. In selenium-poor areas, such as New Zealand, the daily intake may be as low as 30 μg per day, while in selenium-rich regions it can be over 200 μg per day. In the United States it ranges from about 50 μg to 150 μg daily. A daily intake of 30–200 μg per day seems to be safe.

Table 45 Selenium satisfactory daily intakes in micrograms

Years	*µg*
1–10	20–25
10–18	40–60
18–50	65–80
Over 50	80–95

Deficiency

Selenium deficiency causes heart and skeletal muscle damage.

An inadequate amount of selenium in the diet may produce degeneration of the cardiac muscle and general skeletal muscle pain. The heart damage can be fatal. Selenium deficiency is also believed to cause low fertility in men. Whether selenium lack predisposes to cancer is as yet uncertain.

Toxicity

Increasing the normal dietary intake of selenium by an extra 150 µg or more (by using supplements) seems to have caused nausea, vomiting, nail defects and hair loss. The total intake of dietary selenium plus supplement selenium should not exceed the toxic level of about 800 µg per day.

Chapter 45

Zinc

Zinc is an essential dietary item. It is needed for many cellular enzymes involved in the metabolism of carbohydrate, protein and fat and for the synthesis of protein and DNA. There are high concentrations of it in the male reproductive system and also in parts of the eye. The total body content of zinc in an adult is 2–3 g, which is about half that of body iron. As almost all the zinc in the body is in use a regular daily intake is needed, especially during growth and tissue healing. There is much zinc in the secretions entering the intestine and zinc deficiency occurs rapidly in severe diarrhoea.

Sources

Rich sources of zinc are meat, poultry, liver, wholegrain cereals, pulses, herring and shellfish. Most animal protein has about 15 mg of zinc per 100 g. Vegetables and fruit are poor in zinc and their phytate content diminishes zinc absorption into the blood, as does a high calcium intake. Despite the marked differences in zinc content between animal and plant foods, vegetarians do not seem prone to zinc deficiency.

Requirements

The zinc requirements of children are based on their energy intakes, being about 5 mg per 1000 kcal per day. Many children do not get this amount and they would probably show a growth spurt if their intake of the metal were increased. For non-pregnant adults about 15 mg per day are needed, of which about 10 per cent enters the blood. For pregnant women, to ensure optimal growth of the developing baby and to provide it with a good store of zinc, about 20 mg per day are recommended for the last 25 weeks of pregnancy. During lactation zinc intake should be increased to about 25 mg per day.

The average daily intake of zinc in the United Kingdom is only 10 mg per day, which should be increased.

Zinc is relatively non-toxic but massive doses can be harmful. Long-term dosage with moderate but unneeded doses should be avoided because the effect of such a regime is unknown.

Deficiency

> Many children and adults need an increase in dietary zinc.

Studies in the United States have shown that mild zinc deficiency is common in all sections of the population and children given extra zinc have had a spurt in growth. Zinc-deficient children tend to have more infections, grow less well and may have diarrhoea. If these children are fed an increased diet without added zinc they become obese because protein synthesis is diminished and the extra energy consumed is stored as fat.

In adults, zinc deficiency produces loss of appetite, diarrhoea (which causes zinc loss in the faeces), diminished taste, anaemia, skin erosions and scalp hair loss. Zinc lack is made worse by alcoholism, liver disease, kidney disease, rheumatoid arthritis and surgical operations.

If extra iron and zinc are both needed, the supplements should not be given at the same time because the commonly used iron preparations greatly lower the absorption of zinc into the blood. Instead, the iron should be given on one day and the zinc the next day and so on.

On an ordinary UK diet there is not enough phytate to make much difference to zinc absorption but food faddists eating very large quantities of bran can induce zinc deficiency.

At the time of writing, four controlled trials have suggested that zinc may be beneficial in the treatment of the common cold but four other trials have found no benefit.

Chapter 46

Copper and molybdenum

Copper

There are about 100–150 mg of copper in the adult, the highest concentrations being in the liver and the central nervous system, with the kidneys, heart and hair also being copper-rich. The metal is essential for optimal absorption into the blood of iron, for normal red blood cell production, for bone formation and for some oxidation-reduction reactions in the cells, where copper is found in many enzymes.

An adequate copper intake is necessary for normal blood formation.

About half the ingested copper is absorbed into the blood and an equal amount leaves the body via the bile and enters the intestinal lumen, appearing in due course in the faeces. There is very little copper in the urine. The absorption of dietary copper is diminished by a high intake of calcium.

Sources

Copper is found in virtually all unprocessed foods, with the amount depending on the copper content of the soil in the region where the food was grown. The best sources are liver, dark-green vegetables and wholegrain cereals. Milk, milk products, meat, poultry, fish and fruit are poor in copper. Virtually all mixed diets in the United Kingdom provide about 1–3 mg of copper per day, which is about the daily adult requirement. A daily intake of about 2 mg of the metal will keep an adult in copper balance. Children need about 0.05–0.10 mg of copper per kilogram of body weight per day. Pregnant and lactating women require about 0.5 mg per day more than other adults. The full-term newborn has a store of copper in the liver to satisfy its needs until weaning. Premature babies do not have such a store and may become anaemic unless a copper-rich formula is fed.

High intakes of copper are toxic, leading to anaemia, jaundice, coma and death.

Deficiency

It is very unlikely that a dietary deficiency of copper occurs in the United Kingdom at the present time because food from many diverse regions is available. Where only locally grown food is eaten, copper deficiency may occur if the soil is copper-poor. If there is a lack of the metal in the diet the absorption and metabolism of dietary iron is inadequate and an anaemia develops. In infants, but not in adults, the bones may lose substantial amounts of calcium.

Molybdenum

Molybdenum occurs in all plants and animals. It is a constituent of intracellular enzymes, particularly in the liver and kidneys. It has an interesting relationship with dietary copper: when copper intake is high it tends to flush out body molybdenum and a high intake of molybdenum flushes out copper. If too much molybdenum is taken, the loss of copper from the body can cause anaemia because the red blood cells fail to mature.

The best sources of molybdenum are wholegrain cereals, meat and beans (up to 5 parts per million); fruit and vegetables are poor sources (less than 1 part per million). The recommended daily intake of molybdenum at all ages is about 2 μg per kilogram of body weight. Deficiency of dietary molybdenum is exceptionally rare. The average daily dietary molybdenum intake in the United Kingdom is about 0.5–2 mg, about four times that of the United States.

Chapter 47

Magnesium

The behaviour of magnesium in the body is in many ways similar to that of calcium. Its absorption into the blood from the intestine and its storage in bone mimics that of calcium and alteration in magnesium intake affects the metabolism of calcium as well as that of potassium and sodium. It is the most abundant metal ion in cells after potassium and is essential for the maintenance of the membranes of the mitochondrial bodies within cells, the places where reactions occur releasing energy. It is also in many enzymes, especially those involved in producing adenosine tri-phosphate (ATP), which stores energy. Magnesium is required for the synthesis and stability of DNA and for the synthesis of protein. In addition, it plays an important role in the transmission of impulses from nerve to muscle and also in muscle relaxation. In the kidney, it diminishes the production of calcium oxalate stones.

There are 20–30 g of magnesium in the adult, about half in bones and about a quarter in the skeletal muscles.

Sources

Foods vary widely in their magnesium content, the best sources being dark-green vegetables, the magnesium being in the green pigment called chlorophyll, which gives these plants their colour. Offal and skeletal muscle are also good sources, as are wholegrain cereals, nuts and chocolate. Milk, dairy products and fruit are relatively poor sources. The absorption of magnesium into the blood is reduced by calcium, much fat, phosphates, phytates and alcohol.

The average mixed diet in the United Kingdom provides about 200–500 mg of magnesium per day, of which about one-third will enter the blood, the rest being lost in the faeces.

Requirements

Balance studies on magnesium suggest that infants need 40–70 mg per day, increasing to about 250 mg per day by the age of 10 years. Normal adults require about 300–400 mg per day, with that value rising to about 450 mg per day during pregnancy and lactation. These values can be altered considerably if the intakes of calcium, phosphate, potassium and protein are particularly high or low.

Deficiency

Mild magnesium deficiency may occur in various diseases and in chronic alcoholism but is rare in healthy people on a mixed diet. Repeated vomiting and severe diarrhoea, especially in children, may induce magnesium deficiency, as may prolonged use of diuretics.

Lack of dietary magnesium may cause loss of appetite, nausea, general weakness, muscular tremors and convulsions.

Chapter 48

Aluminium, cadmium, cobalt, germanium, manganese, nickel, silicon, strontium, sulphur and tin

Aluminium

Aluminium is the third most abundant element in the earth's crust and traces of it are found in virtually every food. The main sources of aluminium are some antacids, some medicinal products, foods in which aluminium-based additives are used, tap-water from which aluminium salts added during processing are not completely removed and aluminium cooking utensils used for acidic foods such as jams, pickles, vinegars, most fruits and rhubarb. Utensils lined with a non-stick surface do not liberate aluminium. Aluminium foil and trays used for food do not liberate aluminium unless the food is acidic. Aluminium cans for acidic drinks should be lined with a protective layer to prevent leaching out of the metal. Some baby foods may have a relatively high aluminium content from the cows' milk or soya used in their preparation.

> Using aluminium kitchenware for acidic foods (vinegar; fruit juices) can greatly increase aluminium intake.

From ordinary dietary sources the daily intake of aluminium is probably about 6 mg; the intake from the non-dietary sources can be many times greater. Very little of ingested aluminium is absorbed into the blood and most of that is excreted in the urine, but it is possible that some people are less able to eliminate the absorbed metal.

Toxicity

People living in areas where there is a high level of aluminium in the drinking-water have been found to be at greater risk of brain damage leading to a progressive dementia with memory loss as a major symptom. The condition is known as Alzheimer's disease and aluminium is found deposited in a characteristic form in the brain. On the other hand, people taking aluminium-based antacids on a regular basis have not been found to be prone to this disease. The relationship, if any, between dietary aluminium and Alzheimer's disease remains unclear but as aluminium does not seem to be a necessary nutrient it is probably prudent to keep its consumption to a minimum.

Cadmium

Very small amounts of cadmium are present in human tissues and the quantity increases slowly with age, suggesting that the metal is merely a contaminant. Yet that may not be so because human kidneys contain a cadmium-containing protein, which suggests the possibility that cadmium may have a biological role. The average amount taken up into the blood each day by an adult is about 25 μg, of which only about 10 μg are excreted via the urine and the faeces.

Excess body cadmium produces anaemia, damage to the kidneys and damage to the lungs (emphysema), which together may prove fatal. Cadmium poisoning from food is extremely rare. Supplements containing cadmium should never be taken.

Cobalt

The only form in which cobalt is used by the body is as vitamin B_{12}, all of which is produced originally by micro-organisms. Apart from this vitamin there is no known nutritional need for cobalt. Any cobalt in the diet is excreted mainly via the urine, with the rest leaving the body in the faeces.

If excess cobalt is taken in the diet, as in some beers, where it has been used to stabilise the foam, it may greatly increase the production of red blood cells, producing a disease called polycythaemia. There may also be damage to the heart muscle, to the thyroid gland and to nerves.

There is never any need to take cobalt supplements.

Germanium

Germanium is in the diet in trace amounts and as such does no known harm, being easily excreted by the kidneys. There is no evidence that it is required nutritionally or that it conveys any health benefits. Despite this, it has been used in health supplements for a variety of diseases and large doses, 50–250 mg per day for many months, have been taken. Under these conditions it causes damage to the kidneys, the heart muscle and the skeletal muscles. Some people have died from germanium poisoning. Items containing germanium should be avoided.

Manganese

Manganese is known to be essential in animal nutrition and because it is involved in activating many enzymes in human cells it is probably essential in human nutrition. A person living on an experimental diet showed poor blood

Table 46 Manganese satisfactory intakes in milligrams

Years	*mg*
1–3	1.0–1.5
3–6	1.5–2.0
6–10	2.0–3.0
Over 10	3.0–5.0

clotting, weight loss and a reduced blood cholesterol level: these abnormalities resolved when manganese was added to the diet.

Manganese occurs in many foods, especially in tea, which in the United Kingdom provides about half the daily intake of 4–5 mg. Other sources are wholegrain cereals, leafy vegetables, nuts and seeds. Animal and dairy products are relatively poor sources.

Provisional intakes are given in Table 46. Ingestion of large doses of manganese cause the liver concentration of the metal to rise but there seems to be no ill effect.

Nickel

Nickel is found bound to DNA and RNA and is therefore in every cell. Some is also found bound to protein in the blood. When the iron content of the diet is adequate nickel seems to improve iron's utilization. The best sources are cereals, fruits and vegetables; animal products are generally not good sources. About 15–25 μg of nickel per 1000 kcal seem to be needed each day. Lack of adequate nickel in the diet probably does not occur.

Silicon

Silicon probably plays an essential role in the calcification of bone and the production of connective tissue. Wholegrain cereal products are the best sources. The amount needed each day is not known. Deficiency does not seem to occur.

Strontium

Although widely spread in nature and present in food, especially dairy products, strontium does not seem to be an essential nutrient. It is treated by bones as if it were calcium and it is laid down in bones according to its abundance in the blood. It is easily excreted in the urine.

The importance of strontium rose rapidly when nuclear weapons were exploded in the atmosphere because long-lasting radioactive strontium was then spread over vast areas, contaminating milk production, the cows ingesting either the radioactive dust or getting the radioactive strontium in the grass. One of these radioactive forms of strontium (strontium-90) has a half-life of 28 years, which means that it takes 28 years for the radioactivity to drop to half its original power and another 28 years to fall by half again. Hence radioactive strontium entering the bones of young children would still have a quarter of its radioactivity after 56 years. The damage done by this radioactivity would depend on how much there was in the whole body. The strontium can be partially removed, over a prolonged period, by feeding a diet high in calcium, because this element can replace the strontium.

Strontium supplementation is never needed.

Sulphur

Sulphur is present in every cell, the highest concentrations being in hair, skin and nails. It produces enough sulphur dioxide when these tissues are burnt to give a characteristic smell. All proteins contain sulphur because it is part of the amino acids methionine, cystine and cysteine. There is sulphur in the vitamins thiamin, pantothenic acid and biotin. When the liver detoxifies substances, sulphur is often involved in the process. There is no recommended intake for sulphur and human deficiency of this element does not occur.

Tin

There seems to be no tin in the tissues of the newborn; traces of the metal are found in the normal postnatal period. Its role in human nutrition is not known but in animals experimental deficiency leads to poor growth. Deficiency in humans does not occur. Supplements containing tin are never needed.

Index